INTEGRATION OF GEOPHYSICAL TECHNOLOGIES IN THE PETROLEUM INDUSTRY

The most utilised technique for exploring the earth's subsurface for petroleum is reflection seismology. However, a sole focus on reflection seismology often misses opportunities to integrate other geophysical techniques such as gravity, magnetic, resistivity and other seismicity techniques, which have tended to be used in isolation and by specialist teams. There is now growing appreciation that these technologies used *in combination with* reflection seismology can produce more accurate images of the subsurface. This book describes how these different field techniques can be used individually and in combination with each other and with seismic reflection data. World leading experts present chapters covering different techniques and describe when, where and how to apply them to improve petroleum exploration and production. It also explores the use of such techniques in monitoring CO_2 storage reservoirs. Including case studies throughout, it will be an invaluable resource for petroleum industry professionals, advanced students and researchers.

HAMISH WILSON started his career with BP before forming his own consulting company Paras Ltd, focusing on exploration strategy and process. Wilson's key specialism is in helping companies develop and implement exploration-led growth strategies. He has worked at a senior executive level for most of the world's leading exploration-led oil companies. He has now formed BluEnergy focused on carbon capture and storage.

KEITH NUNN is a geophysicist with more than 40 years of experience in oil and gas exploration. He is an independent consultant at Nunngeo Consulting Ltd and has worked for a number of independent oil companies and government oil organisations. He worked for BP for 25 years, mainly in technical leadership roles in seismic processing, analysis and acquisition teams, and finished his career as Distinguished Adviser in Geophysics with responsibility for BP's global seismic activities. Prior to joining BP he was a lecturer in geophysics at the University of Birmingham for 10 years. His research at that time included topics from shallow reflection seismology to large-scale crustal seismic studies. He organized and participated in several crustal seismic projects in the United Kingdom and Europe. He has been involved with exploration and production in most of the major petroleum basins of the world.

MATT LUHESHI is a geoscientist with more than 40 years of experience in oil and gas exploration. He works as an independent consultant at Leptis E&P Ltd and has advised a number of major and independent oil companies as well as government oil organisations.

He previously worked at BP, where his roles included project management and leadership positions exploring new opportunity access, technical evaluations and technology development. His global experience includes many of the major petroleum basins of the world including the Middle East, North-West Europe, North America, North Africa, South-East Asia and East Siberia.

INTEGRATION OF GEOPHYSICAL TECHNOLOGIES IN THE PETROLEUM INDUSTRY

Edited by

HAMISH WILSON

KEITH NUNN

MATT LUHESHI

CAMBRIDGE
UNIVERSITY PRESS

CAMBRIDGE
UNIVERSITY PRESS

University Printing House, Cambridge CB2 8BS, United Kingdom

One Liberty Plaza, 20th Floor, New York, NY 10006, USA

477 Williamstown Road, Port Melbourne, VIC 3207, Australia

314–321, 3rd Floor, Plot 3, Splendor Forum, Jasola District Centre, New Delhi – 110025, India

103 Penang Road, #05–06/07, Visioncrest Commercial, Singapore 238467

Cambridge University Press is part of the University of Cambridge.

It furthers the University's mission by disseminating knowledge in the pursuit of education, learning, and research at the highest international levels of excellence.

www.cambridge.org
Information on this title: www.cambridge.org/9781108842884
DOI: 10.1017/9781108913256

First published 2021

Printed in the United Kingdom by TJ Books Limited, Padstow Cornwall

A catalogue record for this publication is available from the British Library.

ISBN 978-1-108-84288-4 Hardback

Contents

List of Contributors   *page* vii

1   Introduction   1
    HAMISH WILSON

2   The Hydrocarbon Exploration Process   7
    HAMISH WILSON

3   Crustal Seismic Studies   39
    JANNIS MAKRIS AND KEITH NUNN

4   Gravity and Magnetics   71
    ALAN B. REID

5   Full Tensor Gradiometry   124
    MATT LUHESHI

6   Marine Electromagnetic Methods   169
    LUCY M. MACGREGOR

7   Ocean Bottom Marine Seismic Methods   220
    IAN JACK

8   Microseismic Technology   272
    PETER M. DUNCAN

9   A Road Map for Subsurface De-risking   318
     HAMISH WILSON

Glossary   338
Index   341

Contributors

Peter M. Duncan was founding President of MicroSeismic, Inc., a Houston-based geophysical service company where he now serves as President and CEO. He holds a PhD in geophysics from the University of Toronto. He began his career as an exploration geophysicist with Shell Canada before joining Digicon Geophysical, first in Calgary and then in Houston. In 1987 he helped Digicon found ExploiTech Inc., an exploration and production consultancy. He was named President of ExploiTech when it became a subsidiary of Landmark Graphics in 1989. In 1992 he was one of three founders of 3DX Technologies Inc., an independent oil and gas exploration company, where he served as Vice President and Chief Geophysicist. Duncan was 2003–4 President of the Society of Exploration Geophysicists (SEG). Duncan was the Fall 2008 SEG/AAPG Distinguished Lecturer speaking on the subject of passive seismic at 45 venues around the world. He is an Honorary Member of the Society of Exploration Geophysicists (SEG), the Canadian Society of Exploration Geophysicists (CSEG), the Geophysical Society of Houston (GSH) and the European Association of Geoscientists and Engineers (EAGE). He received the Enterprise Champion Award from the *Houston Business Journal* in 2010 and the World Oil Innovative Thinker Award in 2011. In 2013 he was named Technology Entrepreneur of the Year for the Gulf Coast Region and National Energy, CleanTech and Natural Resources Entrepreneur of the Year in the Ernst and Young (EY) Entrepreneur of the Year Program. In 2014 he was awarded the Virgil Kauffman Gold Medal by the SEG.

Ian Jack is a Physics graduate from the University of St Andrews, Scotland. He began his geophysical career in 1968 as a field engineer, and subsequently as a seismic processor, before moving to software development in Dallas, Texas. He joined BP Exploration in London in 1978 and became manager of their geophysical operations in 1982. Other assignments included subsurface R&D manager, and eventually as BP's so-called distinguished advisor on geophysics he was responsible for new technology and subsurface technology strategy during and subsequent to the Amoco and Arco acquisitions. He has championed permanent ocean bottom reservoir monitoring and was the instigator of BP's first 'Life of Field Seismic' project at Valhall in Norway. He is an honorary member of SEG and is a member of EAGE and the Petroleum Exploration Society of Great Britain. He was Vice-President of the SEG in 1992–93 (the first European to hold this role). He chaired the Advisory Committee of the EAGE and has served as a member of the UK's Earth Science & Technology Board of NERC. He was the SEG's inaugural Distinguished Instructor in 1998 with the course entitled 'Time Lapse Seismic in Reservoir Management' which was taught to more than 2,000 geoscientists worldwide during that year. He continued to teach an updated course.

Matt Luheshi is a geoscientist by profession with more than 40 years of experience in oil and gas exploration (mostly with BP). His experience included project management and leadership positions including new opportunity access, technical evaluations and technology development. His global experience encompasses most of the major petroleum basins of the world (including the Middle East, North-West Europe, North America, North Africa, South-East Asia and East Siberia). Following retirement from BP he has worked as an independent consultant (Director at Leptis E&P Ltd) with a number of major and independent oil companies as well as government oil organisations.

Lucy M. MacGregor is Chief Technology Officer at Edinburgh Geoscience Advisors Ltd. She is an accomplished research scientist with more than 20 years of experience in geophysics, multi-physics reservoir characterization and data analysis. She has expertise in innovative workflow and algorithm development to solve technical challenges, as well as significant international experience, both technical and commercial. MacGregor has extensive experience in all legal aspects of intellectual property portfolio management and strategy. She is skilled in managing technical teams made up of specialists with a diverse range of skills to ensure successful project outcomes. MacGregor has 25 years of experience in marine electromagnetic surveying and multiphysics analysis and its application to the detection and characterisation of fluids in the earth. She co-founded OHM in 2002 to commercialize marine electromagnetics in the hydrocarbon industry and spent 16 years as Chief Technology Officer. She was the 2016 SEG Honorary Lecturer for Europe and will serve as the 2021 SEG Distinguished Lecturer.

Jannis Makris was a Professor of Geophysics at the Institute of Geophysics, Hamburg University, from 1978 to 2002, Professor Emeritus from 2002 and

Managing Director of GeoPro GmbH Hamburg. He is currently Managing Director of Geosyn Geophysics. He was an establishing member of the Centre of Marine Research and Climatology at the University of Hamburg and a member of the Board of Directors for more than 20 years. Throughout his long career he has developed marine seismographs for Wide Aperture Reflection/Refraction Profiling (WARRP) for passive and active seismic studies and developed techniques in gravity, magnetics and seismicity. He has organized and coordinated more than 100 geophysical projects in many parts of the world, both onshore and offshore, and published more than 150 papers on crustal seismology, gravity and magnetics, as well as microseismicity. He was Vice President of the UNESCO IOC Commission for the preparation and publication of the geophysical overlay sheets of all the countries in the Mediterranean Sea. For more than a decade he was President and Vice President of the Geophysics and Geology section of Community Integrated Earth System Model (CIESM). He also acted as coordinator of the JOULE Geophysical Project of the European Commission for 5 years. For 20 years he was a member of the International Gravity Commission of the International Union of Geodesy and Geophysics (IUGG).

Keith Nunn is a geophysicist by profession with more than 40 years of experience in oil and gas exploration (mostly with BP). Prior to joining BP he was a lecturer in geophysics at the University of Birmingham for 10 years, during which his research interests included a wide range of seismic topics from shallow reflection seismology to large-scale crustal seismic studies; he organised and participated in several crustal seismic projects in the United Kingdom and other European countries and was a member of the Royal Society Explosion Seismology Working Group. He joined BP as a research geophysicist and spent 25 years mainly in technical leadership roles leading seismic processing, analysis and acquisition teams, and at the end of his tenure served as Distinguished Adviser Geophysics with responsibility for BP's global seismic activities. He has been involved with exploration and production in most of the major petroleum basins of the world. Following retirement from BP he has worked as an independent consultant (Managing Director at Nunngeo Consulting Ltd) for a number of independent oil companies as well as government oil organisations.

Alan B. Reid is a director of Reid Geophysics Ltd and honorary research fellow, School of Earth & Environment, University of Leeds. He specialises in potential field surveys (mainly airborne) used in mineral and petroleum exploration, specifically: designing geophysical surveys to solve geological problems; QC of airborne surveys; advanced processing of geophysical data to display geological structure to best advantage; and interpreting such enhanced data in geologically reasonable ways, consistent with other geological or geophysical information.

Hamish Wilson started his career with BP before forming his own consulting company focusing on exploration strategy and process. Hamish's key specialism is in helping companies develop and implement exploration led growth strategies.

Integrated geoscience focussed on articulating risk and value, is central to Hamish's approach to exploration. He has implemented and peer reviewed exploration strategy and processes in most of the world's major oil companies. An understanding of how oil companies make decisions has led to projects advising governments on attracting exploration investments. Hamish has led influential projects in the UK, Ireland and Canada that have had major impacts on subsequent exploration activity in these countries.

Access to data and the application of appropriate software has been an important part of the exploration and production process. To support the exploration consulting service, Hamish has also previously assembled and led a team of leading experts in the field of petro-technical information technology.

Alongside Hamish's career in oil and gas, he has led a renewable energy start up through to investment funding. The insights gained from advising on oil company capital allocation processes, combined with a deep understanding of renewable energy risk and reward profiles, give Hamish unique perspective on how to succeed in the transition to a low carbon future. He has formed a second consulting company focussed on helping oil companies with the energy transition.

Hamish has presented at most of the major E&P conferences and is a well-regarded opinion leader in his field. He has also served as president of the Petroleum Exploration Society of Great Britain, a charity that represents the exploration community in the UK.

1

Introduction

HAMISH WILSON

This is a 'go-to' guide for decision makers and professional explorers – of the suitability and applicability of each technology in a variety of geological settings. The guide looks at specific hurdles for de-risking the subsurface, such as hard chalk at seabed, salt and basalt and indicates what the optimal technology combinations are best to overcome them. It also gives an indicative cost required for each technology broadly relative to that of a conventional seismic survey.

The search for both minerals and oil and gas is founded on the ability to understand the subsurface and be able to predict the location of commercial bodies of ore and hydrocarbons; and as we move into the energy transition era, the ability to monitor the behaviour of carbon dioxide stored in underground reservoirs will become more important. Traditionally this skill was the realm of the geologist who mapped the surface rock formations and then extrapolated downwards to create a three-dimensional mental image of the hidden layers of rocks. The realisation that these rocks have differing physical properties led to the development of geophysical measurements to constrain the image of the rock formations deep underground; for example, differences in density led to gravity modelling and differences in resistivity led to the measurements of electrical currents. However, it was the understanding of the acoustic properties of rocks and fluids that was the breakthrough in oil and gas exploration.

Reflection seismic and the ability to create an acoustic image of the subsurface has been the backbone of the hydrocarbon industry and opened up all the offshore continental shelves to oil exploration and development. Continual advances in reflection seismic techniques have led to oil being found in deeper water, underneath salt bodies and in smaller accumulations. This has now become the most important geophysical tool used by exploration and production companies and has been extremely successful in helping to locate and produce oil and gas fields.

But this focus on reflection seismic has been to the detriment of other forms of geophysical measurement. Gravity, magnetic, resistivity and seismicity techniques, collectively called potential field techniques, have tended to be used in isolation and by specialist teams. Each is generally tested and used on a stand-alone basis; there is much reporting and analysis on the physics of each technology and how they should be applied individually.

However, increasingly, there is the appreciation that these technologies used in combination with reflection seismic will provide a better answer, that is, produce a more accurate picture of the subsurface. No single technology can solve the subsurface imaging problem. This is the area that is the focus of the book – how should these geophysical technologies be used, both individually and in combination with other techniques?

The lead authors of the book (Keith Nunn, Matt Luheshi and Hamish Wilson) have assembled a group of industry leading authorities on geophysical technologies, each a master of their respective technology. Each has prepared a chapter on their technology and how the physics works. The key insight and differentiator for the book is the description of how the integration of these technologies with other techniques can enhance our understanding of the subsurface. The authors have worked together to describe how these technologies can be combined to solve specific geoscience problems in subsurface under-standing. This volume describes some of these technologies and provides guidelines as to when and where they should be deployed.

The ultimate objective of the book is to present an evidence-based route map of when, where and how to apply these non-conventional geophysical technologies to reduce the exploration and production uncertainty, for conventional oil and gas activity, and to monitor carbon dioxide in reservoirs. Given the specialist nature of these techniques and the fact that they require relatively uncommon expertise, there is a need for a guide for the more generalist practitioner in the field. The route map presented here is a practical technology selection guideline for organisations and individuals working the subsurface. The final chapter of the book describes a 'roadmap to subsurface de-risking' which demonstrates where these specialist tools sit within the tool kit that geoscientists rely on in their day-to-day work.

So in addition to providing an in depth description of each technology, at its heart, this book is a description of when and how to use each technology.

1.1 Introduction to the Technologies of Hydrocarbon Exploration

The oil and gas industry uses technology to create accurate descriptions of the geological architecture of the subsurface. Chapter 2 sets the context for the book and describes the overall process for subsurface analysis in exploration and production (E&P) and introduces where each technology should be used and the problem to be solved. As a company evaluates a given E&P opportunity, or project, money is spent on acquiring and interpreting geophysical measurements (generally reflection seismic) to reduce the uncertainty in our understanding of the subsurface. The focus of the work is to predict the lithologies in the subsurface through measuring their response to various forms of geophysical 'stimulation' or their inherent physical characteristics.

The ability of geophysicists to ascribe physical properties to a three-dimensional geological 'model', to calculate a synthetic geophysical response and iteratively match this to measured data (seismic, magnetic, etc.), has transformed the exploration process. The 'model' closely resembles a three-dimensional picture of the earth. But the 'model' is just that, a picture that can create a geophysical response similar to that measured in the field! As a result, explorationists use different techniques to validate, reconcile and improve models before finally committing to expensive drilling. Validating such models using a range of geophysical measures can transform the level of confidence in a geological interpretation.

The resultant geological model is tested through drilling a well which then calibrates the geophysical measurements against the actual lithologies penetrated. The 'rock physics' of the lithologies penetrated can be calculated and their geophysical characteristics defined along the wellbore. This information allows the model to be 'tied to the well' and updated. The geophysical model is re-created with the new calibration information and the geological prediction refined before the next well is drilled. The process continues through multiple iterations as information is acquired through the exploration process until ultimately, the model approaches the low-risk geological fact.

Today's process is still predominantly reliant on seismic data. Yet there is a suite of other, geophysics-based technologies that can augment the power of seismic to better understand the subsurface. The core of the book is a review of these geophysical technologies, with the premise that the integrated application of these methods fundamentally enhances our ability to reduce uncertainty in the subsurface. This improved ability to characterise subsurface systems is a real game-changer with applications from exploration through to development and production, off- and onshore.

The ability of reflection seismic technology to image the spatial and stratigraphic distribution of plays and prospects is unsurpassed by any other technique. Other methods such as potential field (gravity and magnetics) tend to be used in basin-wide analysis when seismic is either not available, is too expensive or too difficult to acquire or is compromised because of quality issues. Knowing when and how to use each technology is founded on understanding the underlying physics and what rock or fluid property is being measured (density, resistivity, acoustic impedance, etc.) and its resolution. Clearly these technologies measure changes in these properties. Thus, any use of these technologies has to begin with a view as to what you are looking for and the physical properties of the subsurface environment being studied. In this regard Tables 3.1, 4.1, and 4.2 and Figure 6.2 are critical in showing these physical properties of the different rock types.

Geoscientists reach for potential field tools when, for some reason, seismic of an acceptable quality is not available. Hence opportunities for application of potential field methods have tended to fall into one of the following categories:

- Basin identification and definition at the early stages of exploration
- Structural imaging in zones of poor seismic data (sub-salt or sub-basalt)
- Building velocity models to improve seismic imaging
- Characterisation of reservoir properties in areas where seismic gives ambiguous results

The oil and gas industry is continually looking for the 'holy grail' of detecting hydrocarbons before drilling wells. Many remote sensing techniques have been announced as having 'the answer' but have been of questionable technical validity or value. There are now a number of geoscience technologies that are looking increasingly interesting for geoscientists. Advances in acquisition and processing of potential field and electromagnetic data have improved resolution such that they offer a genuine cost-effective ability to either replace blanket 2D or 3D seismic in onshore frontier basins, or to resolve seismic ambiguity. Acquisition at a fraction of the cost of seismic further adds to the attractiveness of these technologies.

No technology is the 'silver bullet' that proves the presence or absence of hydrocarbons or carbon dioxide; however, if used in the 'right' geological conditions and in combination with other techniques, these technologies can reduce risk and uncertainty. The technology landscape is continually evolving with a large and growing evidence base that is available to test the effectiveness of individual technologies. This evidence base allows rational choices to be made as to when and how to use each technology, and which combinations of technologies produce the best results.

We focus on geophysical technologies and have chosen those that have been proven to work and should be in the mainstream of exploration and production. We have deliberately excluded the more fringe technologies, which in our view are either unproven or lack sufficient evidence of success at this stage to recommend them for frequent use.

1.2 Overview

We start the volume with a description of the overall E&P process and basin analysis in Chapter 2. We outline the problem to be solved at each stage and the ranges of geophysical technologies that can be used. We look at what specific questions or issues the technology is being required to solve and at what stage of the E&P value chain is the question being asked. Starting from basin scale issues that need to be addressed by exploration managers during strategic decision-making about what provinces and plays to enter, through block screening, the high-ranking of prospects and evaluation of wells, with the requirement for finer and finer granularity of data and then onto drilling, development and production. We conclude with a description of the application to subsurface carbon sequestration and the need to monitor the behaviour of carbon dioxide in underground reservoirs. At each stage, we discuss the integrated application of these technologies in relation to the workflow and solutions. This chapter leads into the detailed description of each technology.

The core of the volume consists of six technology-focused chapters (Chapters 3–8):

- Crustal seismic studies
- Gravity and magnetics
- Full Tensor Gradiometry
- Marine electromagnetics (controlled-source electromagnetic and magnetotelluric)
- Ocean bottom nodes
- Microseismic and induced seismicity

The technologies are described in order of decreasing scale as illustrated on Figure 1.1 below. Crustal seismic can identify features at the 100-kilometre scale, while microseismic is used to identify fractures at the 10s of metres scales.

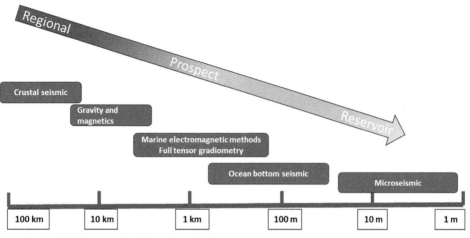

Figure 1.1 The application of technologies and scale.

Each chapter is structured to address the following topics:

- The application and integration of the technology in basin analysis
- History of the technology
- Technical description and basic principles
- How best to apply the technology
- Case examples

In each chapter we have focussed on a discussion of the efficacy of the technology and how it should be applied. We have not reviewed or commented on the companies or vendors who can provide these technologies. Each technology chapter is a stand-alone document.

The concluding piece of our review is the provision of a 'roadmap to subsurface de-risking'. This is presented in Chapter 9. We discuss the optimal way to combine these technologies to reduce uncertainty and risk in exploration and, in some cases, development, production and carbon sequestration ventures.

The book concludes with a glossary and a discussion on units.

2

The Hydrocarbon Exploration Process

HAMISH WILSON

2.1 Introduction

The journey to discovering and producing commercial oil and gas starts with an idea in a geologist's mind. This idea will be a conceptual model which explains when and where oil is generated in a source rock, how it is expelled to migrate into reservoirs and finally, where it is trapped. At the outset, this is just an idea. Through effort and a logical process this idea is turned into an opportunity for an oil company, which then spends money and drills exploration wells. Some ideas will be successful: these become projects that attract further investment through to the construction of facilities which then produce oil and gas.

The continued viability of oil and gas as an energy source is dependent on mitigating the carbon impact of using the product. Thus, carbon capture, usage and sequestration (CCUS) is becoming a critical element of the oil industry's future. This involves the search for subsurface storage sites, potentially depleted oil and gas reservoirs or saline aquifers, close to carbon capture hubs. Thus the application of techniques for finding and producing hydrocarbons is being turned towards understanding how to store and manage carbon dioxide in subsurface reservoirs. As with traditional oil and gas exploration, geophysics makes an important contribution to this effort.

The chapter starts with a description of the exploration process: turning an idea, which by definition is uncertain, into an investable opportunity. In overview, the exploration and production (E&P) process is focussed on building a geological model of the subsurface which predicts the presence of hydrocarbons, and through a process of investment, reduces the uncertainty of the model so that the risk of project failure is acceptable.

We describe the staged approach to exploring for, and producing, oil and gas. First, explorers need to be able to screen basins and find potentially prospective hydrocarbon provinces. Following this regional screening, they then need to identify specific plays that may contain the elements for a working petroleum system (reservoir, source rocks and seal). Then, following a successful exploration programme that identified hydrocarbons, the next stage appraises the scale and productive characteristics of the discovery to enable the design of an effective, economic development. The traditional final stage is the production of the discovered hydrocarbons, where this is commercially attractive. In many basins the economic life of a reservoir is being extended to allow for the sequestration of carbon dioxide as a vital element in our ability to reduce carbon emissions.

We weave the process for subsurface analysis into the story of basin analysis and introduce the concept of reducing uncertainty and risk of loss at every stage. Our understanding of the geological architecture of the earth's crust is based on very limited data, which means that often a number of widely different possible models could fit the data in any given situation. Taking operational decisions in the context of uncertainty is a risky business that can, and very often does, lead to very large financial losses. For this reason, the industry spends a very considerable amount of effort in trying to reduce uncertainty, and hence risk, in understanding the subsurface setting, a factor critical for commercial success.

E&P geoscientists are skilled and experienced in using a standard approach to identifying and evaluating 'play' systems. The approach is based on an overall workflow that follows a set of logical steps: (1) basin screening, (2) play fairway analysis, (3) prospect definition, (4) exploration drilling, (5) appraisal studies and drilling, (6) field development and finally (7) production management. These stages use a variety of geological and geophysical data, which require thorough and integrated interpretation to develop as accurate a picture of the subsurface as possible.

2.2 Exploration Strategy: Where to Explore?

Discovering new hydrocarbons is still at the heart of the global oil economy. All the hydrocarbons in production today have been found through an exploration programme. And with the global drive to reduce emissions, exploration has to focus on low carbon intensity barrels.

For exploration this means searching for 'better barrels of oil', specifically, oil and gas resources that have lower embedded carbon in their production facilities, lower carbon emissions through their production process and low carbon emissions in their transport to their final market. In addition, the resources have to be produced with minimal environmental damage and have a beneficial social impact (i.e. to maximise the contribution to the UN's 17 Sustainable Development Goals

[United Nations, 2015]). The final constraint is that these resources have to be found at lower cost than before.

The industry is continually challenged to maximise the amount of oil found for a given exploration expenditure. The key metric in exploration is the finding cost per barrel, that is, how much does it cost to find a barrel of oil? Clearly the industry does not wish to waste money. But oil and gas exploration is a risky business in that not all wells find hydrocarbons and a failed well (a dry well) is expensive. The work that is done to 'de-risk' a given well and to choose its location is vital. The industry is continually looking for the least costly way to understand the geology to predict a hydrocarbon accumulation and choose a well location. There is always a tension between a company's investment in 'de-risking' a well and drilling the well itself.

This is where technology comes in. The current mainstay technology for exploration and production is reflection seismic. However, as we argue in this book, there are other geophysical technologies that are cheaper to use and, when used in conjunction with reflection seismic, will give a better answer. These technologies become even more important as companies look to reduce their exploration budgets and find more resources with fewer funds.

The obvious challenge for oil companies is where to look to find the oil and gas resources that match these criteria. Figure 2.1 shows a global map of the world's basins (Roberts and Bally, 2012) and is the starting point for many companies'

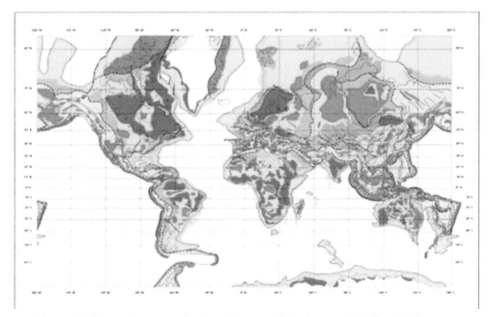

Figure 2.1 The sedimentary basins of the world (Roberts and Bally, 2012).

exploration processes. This shows the wide range of basin types and geographies that are potential exploration targets. The choice of where to look is governed by the company's strategy and the type of exploration it wishes to do.

Exploring in basins that have no discovered hydrocarbons requires different skills, approaches and technology from those for exploring in areas where there are already producing fields and production infrastructure. To guide this thinking, sedimentary basins (and indeed exploration plays[1]) are classed in differing stages of 'maturity' for hydrocarbon extraction (Westwood, personal communication). The industry generally categorises basins into four stages of maturity:

- **Frontier basins** are defined as sedimentary basins that might be of interest to the industry that do not have a commercial oil discovery.
- **Emerging basins** are those in which the hydrocarbon volumes discovered per well continue at the same rate and before the discovery rate starts to decline.
- **Maturing basins** are basins in which the hydrocarbons discovered per well start to decline but before the extended tail which defines the **mature basins**.

There is, of course, an element of subjectivity as to where the latter two categories start and finish. The best way to illustrate the evolving exploration maturity of a basin is using a creaming curve. Figure 2.2 shows typical 'creaming curves' for the basins in the North Sea. This is a graph plotting the cumulative growth of discovered resource (volume of oil and gas commercial discoveries) through time. The graph starts with the first commercial discovery in each basin, noting that in all

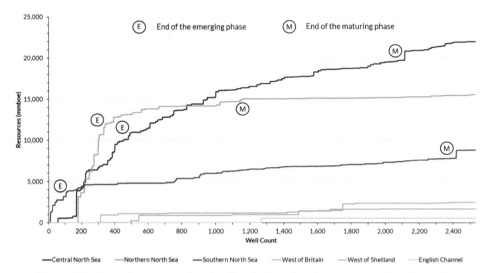

Figure 2.2 Creaming curves for the North Sea showing the transition from the emerging to maturing and mature phases of exploration. (Source: Westwood Global Energy. Used with permission)

these basins there are a large number of failed wells prior to success. The first success is followed by a steep rise in volumes found, as the larger fields are discovered first. Then the basin matures as smaller fields are discovered and produced and exploration focusses on existing production facilities. We also note that success rates decline as the exploration targets get smaller and more difficult to find.

Most of the UK/Norway continental shelf is in the mature phase of exploration, as is the shallow water of the US Gulf of Mexico. While many of the basins off West Africa, such as those in Angola and Ghana, are maturing, the new (2020) discovered basins in Mauritania and Guyana are emerging. Frontier basins such as those in the passive margins offshore Argentina and its conjugate, Namibia, along with the Great Australian Bight are attracting interest. This interest is driven largely by the successes in the deepwater passive margin basins offshore Senegal and Guyana. The intra-cratonic basins have attracted less interest, mainly because any hydrocarbons in these basins have been discovered and produced (onshore USA, Russia, North Africa and basins in the post-Soviet states).

A company's corporate and exploration strategy determines which type of basin and play the company wishes to explore to build a business. Many companies choose to focus on a particular basin and type of exploration – for example, searching for new oil around existing fields or production hubs, or a region. There are also companies that target frontier basins; in recent years these have been the super major oil companies that have the strength of capital to sustain the number of failed wells in frontier basins.

The choice of strategy determines the technology mix and thus choice of particular geophysical technology that adds value to a given exploration programme.

2.3 The Exploration Process

An oil company exploration process can be described as progressing a set of opportunities through a funnel (Figure 2.3). The company looks at a large number of investment options and screens them against their investment criteria, based on the company's strategy. The options that are taken forward are those that fit the overall strategic direction of the company. If the company chooses to invest in an option, then the opportunity moves through the funnel that turns the idea into an oil discovery and on to production to earn money through selling oil or gas. Fortunately for the oil industry the money the oil company makes through selling oil and gas is more than that spent in exploration, appraisal, development and production!

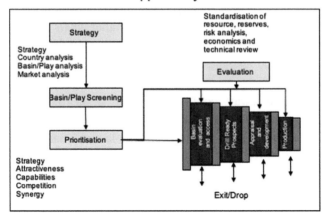

Figure 2.3 The oil company opportunity funnel.

In progressing through the opportunity funnel the company invests in data and analysis to reduce subsurface uncertainty and risk of failure. This is done through the application of technology and drilling wells. Much of the work through the early phases of this opportunity funnel is geared towards the investment decision for building a production facility. The business case for this investment is founded on a description of the subsurface and a projection as to the amount of oil or gas, or both, that can be produced.

Figure 2.3 shows the new opportunity funnel. Note that new opportunities can enter the process at any stage. Clearly if the company is looking for production opportunities, these enter 'later' down the funnel. At each stage of the process the company evaluates the option and decides whether to progress to the next stage or drop it. In this way a company manages a portfolio of investment options at different stages of maturity. The company has choices as to how to 'monetise' that opportunity, either through selling it or progressing on to producing oil and gas and selling that.

Note the importance of screening the opportunities and selecting only those that best fit the company's strategy. The basin/play/opportunity screening is a vital part of an oil company process. In most cases screening involves consideration of a broad variety of factors, generally called 'above ground' and 'below ground'. The above-ground issues include factors such as

- Political risk, country stability and operational safety.
- Route to market for the hydrocarbons. If the opportunity is gas, a key factor would include the size of the local market for the produced gas. For example, there are very small local markets for the large gas discoveries in Mozambique

and Senegal; hence the gas has to be shipped as liquified natural gas (LNG) to the gas market, with an implication for the cost of production.

- The tax regime and the amount the local government takes in terms of royalties (linked through to consideration of political risk and what happens to the oil revenue).
- Local infrastructure, labour and skills.
- Onshore/offshore and water depth.
- Increasing importance of consideration of the UN's 17 Sustainable Development Goals (United Nations, 2015).

The below-ground or subsurface issues are those that most of us geologists are familiar with and include

- The chance of finding hydrocarbons.
- Is it oil or gas?
- How deep is the target reservoir?
- How much oil/gas is there going to be (volume of resource discovered)?
- What production rate is there going to be per well? This controls the number of wells that need to be drilled and thus the overall cost of production.

The most important consideration for the subsurface analysis is the potential volume of hydrocarbons and its risk (i.e. chance of finding commercial hydrocarbons). Generally, oil companies balance below-ground risk against above-ground risk. For example, oil companies are prepared to consider investing in Iraq, where there is a high political risk but minimal subsurface risk. Conversely, companies are investing in Namibia, which is considered politically stable, yet with high subsurface risk: the basins offshore Namibia are still (in 2020) in their frontier stage despite a large number of wells.

It is in this screening phase that geophysical technologies are particularly important as a low-cost way of establishing the presence of a basin and its main tectonic features. These technologies are of particular value in that many are airborne and do not require access to the ground and thus can contribute to establishing the overall basin architecture efficiently and at low cost.

Table 2.1 shows the relevance of each technology considered in this report to the exploration and production phase. It is important to note that conventional reflection seismic becomes cost effective from the exploration phase onwards. Also note that in the exploration phase, there are a range of other technologies, which when used in combination with reflection seismic could reduce subsurface uncertainty. It is this area we turn to next to address the use of these technologies in constructing a geological model of the subsurface that supports the process from exploration through to production.

Table 2.1 *Geophysical technologies and the exploration and production phase*

E&P phase	Basin screening	Access	Exploration	Prospect evaluation	Appraisal	Development	Production
Crustal seismic	■						
Gravity and magnetics	■	■	▨				
Full Tensor Gravity		■	■				
Marine CSEM/MT surveying			■	■	■	▨	▨
Ocean bottom seismic				■	■	■	■
Microseismic and passive seismic							
Conventional reflection seismic	▨	■	■	■	■	■	■

CSEM, controlled-source electromagnetic; E&P, exploration and production; MT, magnetotelluric.
Green indicates a good fit; orange, partial fit; blank, not applicable.

Figure 2.4 The exploration triangle: from regional analysis through to prospect definition.

2.4 The Exploration Triangle and Prospect Analysis

The elements that make up a geological model and how that model is used to support the definition of a prospective drilling location (i.e. a prospect) is best illustrated through the Exploration Triangle (Grant, Milton and Thompson, 1996; Fraser, 2010) shown in Figure 2.4. This shows that prospect analysis, determining a prospect's chance of success (risk) and potential resource volume, depends on fully understanding the regional geology and the basin context. The Exploration Triangle in Figure 2.4 lists the elements that are consolidated together to allow a detailed prospect evaluation.

Notice that gravity and magnetics is explicitly mentioned as being a requirement to help understand the structure of the basin. This is not the only technology of value; the role of technology in building an understanding of the tectonics that drive basin formation is explained in Section 2.5.

Figure 2.5 illustrates the concept of defining the exploration play first before going on to working on the prospect. For most oil companies a play is best illustrated by maps, and the two maps that are most commonly used are the Gross Depositional Environment (GDE) and the Common Risk Segment (CRS) maps. These are prepared for the sequences that define a given play – in general defined by source, reservoir and seal.

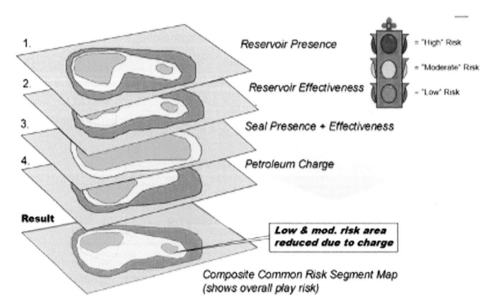

Figure 2.5 Common Risk Segment mapping.

The CRS maps, as illustrated on Figure 2.5, outline areas of similar risk with traffic light colours. Areas coloured in green are those in which that particular element (reservoir, source rock or seal) is certain. Areas coloured in red are those in which we are certain the elements are absent or have a high risk. Amber areas are those in between. CRS maps are produced for each element (source, reservoir and seal and their supporting data) that makes up a prospect. The maps are overlaid and those areas in which the green areas overlap will form the target area. The process was first developed in BP in the 1990s and was published in several papers thereafter. The best description of the methodology can be found in Fraser (2010) and Grant, Milton and Thompson (1996).

It is a very simple concept but the thinking and analysis that go into preparing each map is profound.

2.5 Geophysical Technologies, Subsurface Modelling and Basin Analysis

2.5.1 Basin-Forming Processes

Very simply put, oil and gas are found in sedimentary basins, which are formed by tectonic processes that create depressions. These depressions fill up with sediments carried in by rivers and the sea (occasionally by the wind). The shape of the depression, or basin morphology, is controlled by tectonic processes caused by the movement of continental plates (plate tectonics). As the basin forms and deepens, each phase of tectonism (fault movements) changes the processes in which sediments fill up the basin. This in turn changes the lithology of the sediments. The complete package of sediments that are deposited within one phase of tectonism is called a mega-sequence, which is made up of sequences of sediments bounded by smaller geological events.

Understanding the tectonic processes that control basin formation is a critical component in defining the hydrocarbon potential of a given basin. These tectonic processes drive the deposition of the lithologies that make up the source, reservoir, seal and trap that define a given play and prospects within that play. The use of crustal geophysics and gravity/magnetics has made a profound contribution to building our knowledge of how basins form.

Typically, basins form in an extensional tectonic environment. This extension causes the crust to break and rift to create a depression. The rate and extent of rifting either lead to continental break-up and the formation of passive margin basins or cease, to form a failed rift basin (such as the North Sea) or an intra-cratonic basin (e.g. the Mursuk and Kufra basins of southern Libya). These extension basins become involved in compression events and are then incorporated into either foreland basins (e.g. the Alberta Basin of Western Canada), in which the thrust sheets of the mountain front cause either a depression (isostatic depression) into which more sediments are deposited, or thrusted compressional basins (e.g. Papua New Guinea or the Rocky Mountains).

Potential field technologies (gravity and magnetics), along with crustal seismicity, can contribute to understanding the basin morphology of these basin types and assist in defining the major sediment packages. More recently, regional magnetotelluric (MT) studies (passive source electromagnetic [EM] measurements) have been shown to contribute valuable information in this phase.

2.5.2 Crustal Structure in Passive Margins

In the case of passive margins, the original rifting process controls the deposition of sediments through the rifting process (the syn-rift mega-sequence) and the deposition through the drift phase (the post-rift mega-sequence) as the continents drift apart, the earth's crust cools and subsides. The rate of subsidence of the basin floor is related to crustal thinning, which is in turn related to heat flow: the thinner the crust the higher the heat flow.

Both the rate of subsidence and the heat flow are profound controls on source rock deposition and subsequent hydrocarbon generation as the source rock is buried, lithified and heated up. Classically, lacustrine sources rocks are found in the syn-rift phase of rifting. Examples of these are found in Lake Albert in the East African Rift System in the present day. These are the main source rocks for most of the oil found in the rift basins of Indonesia and Malaysia. As extension continues, the rate of subsidence in the basin defines the extent of the early influx of the sea (the early marine transgression) and the nature and longevity of the shallow seas. Restricted marine settings also result in source rocks.

The behaviour of the flanks of the rift controls the rate and type of sediment input; uplift of the basin margins produces rivers and thus 'clastic input'. Stable and low-lying basin margins result in low clastic input, leading to carbonate deposition. In some cases, carbonates and clastic sediments are interwoven, reflecting changing rates of clastic sediment input.

The rate of subsidence is related to the amount of volcanism through rifting; if there is volcanism, then it is a 'hot margin' and the rate of subsidence is low. Most of the prolific hydrocarbon passive margin basins are on 'hot margins' (Levell et al., 2010). Of particular note in this regard are the basins offshore Brazil and Angola. The passive margins of the North Atlantic tend to be 'cold margins' and have had disappointing exploration results to date.

Determining the rifting process and defining where the basin sits on the 'hot–cold' margin spectrum is the domain of deep crustal seismic. This is also used in combination with long-offset two-dimensional (2D) lines to extend the resolution at depth and image crustal structure. The identification of what are called Seaward Dipping Reflectors (SDRs) seen on crustal seismic analysis indicates volcanism at the rift and immediate post-rift phase (White et al., 1987). The reflectors are thought to be caused by basalt lava flows originally deposited on flat surfaces but dip seaward as the crust subsides.

Recent discoveries in basin floor turbidites offshore Guyana and Mauritania-Senegal have stimulated exploration interest into deeper water exploration. From their very earliest conception, it has long been known that turbidites extend a long way offshore. However, the concern about the deep offshore was that the sediment thickness was not sufficient to mature any syn- to early post-rift source rocks. Crustal seismic analysis prompted the realisation of a thinner crust and higher heat flows and has re-appraised the source rock potential in the deep offshore – oil and gas–prone source rocks extend deeper and further offshore than previously thought (Nemčok, 2016), leading to renewed interest in all deepwater passive margins.

In conclusion, the use of crustal seismic analysis determines the conditions in which source rocks might be deposited through the rifting process, from syn-rift through to early post-rift, and the heat flow which drives the thermal maturation of the source rock.

2.5.3 Exploration of Onshore Basins

Reflection seismic is expensive to acquire in onshore basins. This has generally restricted its use to prospect definition. Traditional gravity (ground-based or airborne) has been a fundamental tool in onshore exploration. The ability to determine the major tectonic basin–forming features and resulting sediment packages has been essential in opening up the exploration potential of onshore basins (e.g. the onshore East Africa Lake Edward example described in Chapter 4, Section 4.11.2). MT and other EM methods have also been applied in the onshore setting; however, these approaches are not discussed further in this volume.

2.5.4 Subsurface Modelling: Pre-discovery

Once the basin structure has been defined and its major tectonic features (e.g. controlling faults) have been established, the sediment fill has to be described. Generally, the combination of potential field analysis (traditional gravity/ magnetics, Full Tensor Gradiometry [FTG] and MT surveying) combined with its overall tectonic setting will indicate if the basin has any hydrocarbon potential. Owing to the great extent of previous exploration initiatives, if a basin has hydrocarbon potential it is likely that there will be regional 2D reflection seismic lines.

The integration of potential field data with regional seismic is particularly powerful in establishing the overall sediment structure and major lithology groups. Potential field data will assist with

- Definition of 'basement' – that is, those lithologies that have no hydrocarbon potential and form the basin outline. These are likely to be metamorphic or igneous rocks.
- Identification of granites
- Salt/shale differentiation
- Carbonate/shale differentiation

For onshore basins, field work is essential in determining the major rock types, calibrating the potential field and seismic data and compiling previously collected data.

These analyses provide the information on which to build a model of the major sedimentary packages and their tectonic controls (i.e. the mega-sequence and sequence structure of the basin fill – the tectono stratigraphy). The thesis here is to build as robust a model as possible prior to investing more in data acquisition – either drilling a well or acquiring reflection seismic.

As confidence in the hydrocarbon potential of the basin grows, further reflection seismic both 2D and 3D will be acquired. Prior to drilling, describing the major lithological packages becomes increasingly important as we define the play fairways and their related prospects. In frontier basins, FTG is essential in identifying the major lithology groups. For more mature basins, in which the stratigraphy and its correlation to reflection seismic is well defined, prospect analysis and the identification of valid traps becomes the key concern.

Controlled source electromagnetic surveying (CSEM) is a valuable tool in prospect identification when the major lithology groups have been identified. CSEM measures resistivity differences between lithologies. Salt-water has a low resistivity while hydrocarbons, and in particular oil, have a high resistivity. Therefore, for a given sandstone reservoir, CSEM can be used to identify those traps that are likely to contain oil. However, to be effective the sequence structure of the basin must be understood and the lithologies reasonably well pinned down and correlated with the seismic. Used in this way, CSEM is used to calibrate the seismic image. This further emphasises the point that these technologies are at their most effective when used in combination with other measurements and analysis.

The description of this work programme and the full geoscience analysis alluded to in the exploration triangle in Figure 2.4, is outside the scope of this book. Suffice to say, this work programme will culminate in the identification of a prospect. The company's 'best' prospects (with 'best' defined according to alignment with the company's strategy) will be tested with a well. If the well is dry, then the dry hole information is fed back to recalibrate the geological model and the other prospects in the same play re-risked. If the well is a discovery, then we move onto the next phase of the hydrocarbon exploitation process – appraisal and development.

2.5.5 Subsurface Modelling: Post-discovery

The work done in defining the prospect will have predicted the shape and size of the potential hydrocarbon accumulation. The discovery well has proven that there are hydrocarbons present in the prospect and the extent to which it confirms the pre-drill prediction. The next step in the process is to appraise the discovery, that is, to drill enough wells to check the extent and producibility of the reservoir. One key factor in particularly deepwater fields, where wells are expensive, is the well productivity – the production rate that can be achieved from one well and the amount of reservoir that is 'drained' by that well. These factors determine the size and scope of the production facility (the platform) and the number of production wells that need to be drilled.

The analysis described earlier is concerned with building a geological and thus static model of the subsurface. From a hydrocarbon standpoint this work defines the shape of the container for the oil and gas liquids and gas. In describing the container, the subsurface model will consider aspects such as the

- Porosity distribution in three dimensions (i.e. the identification of reservoir and seals)
- Lateral and vertical continuity between reservoir horizons
- Permeability distribution
- Location of faults and fractures

With the advent of 3D seismic and more computing power, a three-dimensional model of the reservoir can be built which models the porosity and permeability distribution in three dimensions for the complete discovery. The model is built from the 3D seismic 'cube' that is calibrated with the well data to predict the lithology and thus porosity and permeability distribution. This is a static model.

Increasingly the resolution of conventional towed streamer 3D seismic is not sufficient to build this model and ocean bottom nodes are used. In simple terms, this increases the 'pixel' density of the model and its resolution.

To determine the size and scope of the production facility, and the number of production wells, the static model is then used to create a dynamic model of the reservoir. The behaviour of the oil, gas and water is modelled through the life of the field. As a well starts taking oil out of the reservoir so the pressure will drop and the boundary between the oil and the water will move. The understanding of the behaviour of the oil and the 'oil–water contact' is particularly important to maximise the amount of oil that can be recovered from a given field.

Both the static and dynamic models of the field are updated as production wells are drilled and produced through the production life of the field. As a field is produced, so interventions are planned to make sure the maximum oil is recovered. These will include drilling 'injector' wells to support the pressure and to support the 'sweep' of the oil out of the reservoir. These types of activity are generally called 'enhanced oil recovery' (EOR) programmes.

In planning the optimal way to produce a field and the appropriate interventions necessary to increase recovery, predicting where the oil is and how it has moved is very important. This is where ocean bottom nodes are becoming increasingly effective. These allow identical reflection seismic surveys to be acquired at different times (4D surveys). The reflection seismic data from the different surveys can be compared. As the static model is that – static – the only factor that will have changed is the movement of fluid. Thus, 4D ocean bottom surveys (and, sometimes, 4D gravity surveys) are used to help in reservoir management processes.

2.5.6 Microseismic and Fracture Prediction

Classical seismology has long been an important science to locate the location of earthquakes and thus map out tectonic scale faults. The energy source is the natural movement of the earth. Understanding the location and extent of fractures is now a vital tool in oil and gas production, in particular in exploiting unconventional hydrocarbons – shale gas and shale oil. In these reservoirs, the hydrocarbons are trapped as part of the lithology (as distinct from the spaces between the rocks – the porosity, in conventional oil or gas reservoirs). The hydrocarbons are released by the process of 'fracking'. High-pressure fluid is injected into the rock to create fractures. The factures are 'propped open' using 'proppants', which are generally sand. Gas or oil is released into these fractures and flows to the surface.

Microseismology is the science of understanding the extent and behaviour of the fractures generated by the fracking process. In the same way as classical seismology, the fault (or facture) movement caused by fracking creates a seismic event that can be located in three dimensions. So the fracture planes are located along with their subsequent movement as the gas or oil is extracted from the 'reservoir'.

Mapping the extent of the fracture network in three dimensions informs decisions on well spacing (each well will have a fracture 'halo' around it) and therefore the optimal number of wells to exploit a given unconventional resource.

2.5.7 Carbon Sequestration and Storage in Subsurface Reservoirs

Carbon storage in subsurface rock formations is a vital element in our ability to reduce the carbon emissions to the atmosphere. The reservoirs that are the targets for carbon storage sites are generally in basins that have been thoroughly explored for oil and gas and that have an existing production infrastructure (pipelines and production platforms). Conceptually oil and gas reservoirs are obvious targets for carbon dioxide storage and given their production history these tend to be relatively well understood. However, given the location and quality of infrastructure, much of which is more than 60 years old, 'greenfield' saline aquifers are more likely storage locations for which new facilities must be constructed. The challenge for carbon storage is to understand the movement and behaviour of the carbon dioxide as it is injected into these aquifers. In addition, there will be requirements to make sure that the carbon dioxide stays put.

At typical reservoir temperatures and pressures, CO_2 is less dense than water. Thus, displacing water with CO_2 in a reservoir will generate a gravity response. In addition, replacing water with CO_2 will produce a different acoustic and hence seismic response. Hence a combination of conventional seismic, gravity and perhaps EM methods can be used to monitor and manage CO_2 reservoirs.

2.5.8 Crustal Studies

The presence of most of the global sedimentary basins is well known (see Figure 2.1 in Chapter 2). The challenge for determining the economic potential of the basin, from either a hydrocarbon, or mineral exploration standpoint, is to understand the mechanism for basin formation and the major controlling tectonic features. The process for rifting, and subsequent basin subsidence controls, the sedimentary fill and thus lithologies. Similarly, the location and nature of the major tectonic features is important to establish; these features, could be highs or lows, the nature of which controls the surrounding sedimentary sequences.

Most of these major basin forming features can be identified by gravity surveys as is described in (Chapter 4). Gravity is the cheapest and easiest to acquire of these geophysical techniques. Given that we are considering basin scale features, the main constraint for gravity is that any given gravity response can be generated by a number of subsurface architectures (solutions are non-unique because there is an ambiguity between density and depth, which cannot be resolved with gravity data alone). A much better resolution of lithologies is to use gravity in combination with crustal seismic and using the velocities response to constrain the gravity model and vice versa.

The application of crustal seismic and gravity is illustrated in the case study in Section 3.4.3 to determine the crustal profile of the East European Craton, an intra-cratonic basin. The purpose of the study was to understand the nature of the plutonic blocks and the nature of the continent ocean boundary. The interpretation of the plutons was constrained by a gravity survey.

For hydrocarbon exploration, predicting the presence and effectiveness of a source rock is critical. The conditions for depositing and the subsequent thermal behaviour of the source rock are determined by the how the basin was formed and its rate of subsidence. Thus, predicting paleo-water depth through the rifting process is critical. The faster the rate of subsidence through the syn-rift and post-rift mega-sequences, the less likely the conditions for source rock deposition (lacustrine and restricted marine) and preservation will occur. Conversely, a slow rate of subsidence allows the potential for thicker sequences of source rock lithologies to be deposited. This is not the end of the story.

The generation of oil and gas from a source rock is dependent on the length of time spent at a given temperature, that is, the rate of subsidence and the heat flow. Predicting both parameters is considerably aided by crustal seismic. The case study for both the Kenya passive margin (Section 3.4.1) and Nova Scotia passive margin (Section 3.4.2) are excellent examples of the application of crustal seismic to determine the nature of basin forming rifting process and subsequent rate of subsidence.

This chapter describes the history of the technology and its earliest application to recognise the discontinuity between the crust and upper mantle – the Moho. The physics of the technology and how a velocity model of the subsurface is created are explained. The final technology overview is of the instruments used to collect the seismic data, both offshore and onshore, before the chapter concludes with four case studies.

Table 2.2 summarises where the technology is used in the exploration process and what it should be used with to improve the resolution. In this case the application is to basin screening and can be used alongside gravity data to determine the main structural features of a basin.

2.5.9 Gravity and Magnetics

The gravity and magnetic survey methods find most application in frontier exploration. Potential field methods are the cheapest and easiest to acquire of all geophysical techniques. They are extremely cost effective at delineation of basins and determining structural controls on those basins, especially delineating normal faulting within rift basins. Airborne gravity is cheap and effective enough to be the initial phase in any exploration programme, with magnetic sensors added to help identify basement inhomogeneities. In a similar way to crustal seismic techniques, gravity and magnetic information help understand the mechanism for basin formation and the major controlling tectonic features. These, as mentioned previously, control the depositional environments, and hence define the lithologies through the early phases of basin formation.

The critical factor in determining the effectiveness of a gravity/magnetic survey is to have some concept as to the lithologies we are looking to identify. Gravity measures density differences, and magnetic measures magnetic susceptibility and remanence and in this regard the key figures to note are Figures 4.8 and 4.10. Surveys need to be planned with concept for the geological model considered and the density contrasts expected at the outset. For example, a gravity survey might not contribute to understanding a stacked sand shale basin fill but would if intruded by igneous rocks or halite. But even with these sequences there may be non-unique resolutions to a given gravity magnetic model. Thus, the interpretation needs to be constrained by other information (e.g. surface geology and 2D seismic).

Gravity can be used in combination with crustal seismic and using the velocity response to constrain the gravity model to resolve ambiguities. Thus, integrating the techniques where appropriate reduces the uncertainty in lithology prediction.

Magnetics has three functions in early-phase work:

1. Identifying approximate basin morphology
2. Identifying igneous intrusives and extrusives

Table 2.2 *Crustal seismic technology in the exploration and production phase*

E&P phase	Basin screening	Access	Exploration	Prospect evaluation	Appraisal	Development	Production
Crustal seismic	▓						
Gravity and magnetics	▓						
Full tensor gravity							
CSEM							
Ocean bottom seismic							
Microseismic and passive seismic							
Conventional reflection seismic							

CSEM, controlled-source electromagnetic, E&P, exploration and production.
Green indicates a good fit; blank, not applicable.

3. Along with airborne gravity, pin-pointing dense basement ultramafics which influence the gravity response
4. In later-phase work, high-sensitivity surveys can reveal structure in sediments.

This then enables early decisions about cost-effective placement of seismic surveys and other intensive follow-ups.

The application of gravity in frontier exploration is illustrated in the case studies (Chapter 4, Section 4.11). These illustrate how gravity is used to determine the tectonic features and sediment thickness of a basin and thus constrain the basin analysis. Examples are presented from Offshore Guinea (Section 4.11.1), Uganda (Section 4.11.2), the North Sea (Section 4.11.5), the Irish Sea and Western Australia (Section 4.11.6).

Gravity/magnetic techniques are best applied when there are lateral discontinuities in the density of the lithologies to be studied. Therefore, in more mature exploration, gravity and gravity gradient data combine well with seismic data in distinguishing between alternate interpretations and thereby removing ambiguities. High resolution magnetic data offer an effective means of fault connection in conjunction with regional seismic coverage, if shales are present in good quantity. The application to mature exploration and integration with seismic data are illustrated in case studies in the East Irish Sea (Section 4.11.3) and North Sea (Section 4.11.4).

Table 2.3 summarises the where the technology is used in the exploration process and what it should be used with to improve the resolution. Gravity and magnetics are best applied is to basin screening and can be used alongside crustal seismic data to determine the main structural features of a basin. In the access and exploration phases, gravity and magnetics are used in combination with conventional seismic to remove ambiguity in lithology prediction.

2.5.10 Full Tensor Gradiometry

In modern exploration, as discussed in Chapter 4, gravity and magnetics data are often used to help understand regional basin architecture and the mechanism for basin formation. Thus, gravity/magnetic technology is by and large a regional exploration tool and most valuable for imaging the overall fabric of basins. However, there are many examples from the early days of exploration of major oil and gas fields being discovered using gravity and magnetic data alone (e.g. a number of the Middle East fields were discovered in this way).

Modern exploration, however, relies on high-resolution reflection seismic. Seismic allows subsurface imaging with very high resolution, enabling mapping of stratigraphic and structural features in great detail. As described in Chapter 1,

Table 2.3 *Gravity and magnetics and the exploration and production phase*

E&P Phase	Basin screening	Access	Exploration	Prospect evaluation	Appraisal	Development	Production
Crustal seismic	▓						▓
Gravity and magnetics	▓	▓	▓	▓			
Full tensor gravity							
CSEM							
Ocean bottom seismic							
Microseismic and passive seismic							
Conventional reflection seismic							

CSEM, controlled-source electromagnetic, E&P, exploration and production.
Green indicates a good fit; orange, partial fit; blank, not applicable.

explorers use increasingly sophisticated methods for understanding controls on plays and for characterising prospects. Seismic is critical for de-risking exploration, appraisal and development drilling.

Despite the ubiquitous nature of conventional seismic, there are occasions when for some reason seismic of an acceptable coverage and/or quality is not available. This is when potential field tools can offer a solution Examples of the application of traditional gravity, alongside conventional seismic for fault identification and to map shallow sand were described in Section 4.11.

Conventional gravity measures *one component* of the three-component anomalous gravity field (usually the vertical component – but see 'what do we mean by vertical' in Chapter 4). In contrast, full tensor gravity (FTG), however, measures the *derivatives* of all three components in all three directions. This allows a tighter image of the subsurface with a greater granularity.

Note we are still looking for density contrasts, so the density chart in Chapter 4 (Figure 4.8) is key to determining when to use this technology – that is, you must have some concept of the subsurface model at the start of the process. Similarly, as with conventional gravity/magnetics, lateral discontinuities are best imaged. But with three components of measurement (as opposed to vertical), FTG provides greater resolution as to the location of the lateral discontinuities.

FTG is therefore particularly useful in delineating shallow salt bodies, carbonate banks or faults in areas of sparse or poor quality 2D conventional seismic. It is a relatively low-cost way of refining a 2D interpretation to locate potential prospective areas of the basin on which to acquire more expensive 3D data. The most common application is therefore onshore exploration where acquisition of 3D can be prohibitively expensive.

Table 2.4 gives a 'traffic light' summary of where FTG surveys are most useful in exploration to production workflows and the types of structures where it is most applicable.

Table 2.4 *Use of gravity gradiometry in the exploration and production life cycle*

	Basin definition	Exploration prospects	Appraisal	Development	Production
Onshore – low surface relief					
Onshore severe surface relief					
Shallow marine					
Deep marine					

Green indicates a good fit; orange, partial fit; red, no fit or not applicable.

Today application of potential field methods tends to fall into one of the following general categories:

- Mapping regional basin architecture (early stages of exploration); fully described in Chapter 4
- Imaging in zones of poor seismic data (sub-salt/sub-basalt, structurally complex areas)
- As input to the construction of velocity models for used in seismic imaging (seismic migration)

Gravity gradiometry imaging relies fundamentally on lateral density contrasts and is identical to 'conventional' gravity surveying in this respect. This implementation of the gravity methods has two basic advantages due to improved signal-to-noise characteristics and the fact that FTG instruments measure all three components of the gradient of the gravity field simultaneously. The case studies described in this section illustrate these two points. The exploration applications fall into one or a combination of the following categories:

- Direct higher resolution imaging of structural fabric in areas where seismic is compromised (Chapter 5, Section 5.4.1)
- Construction of better constrained seismic velocity fields in service of improving seismic imaging (2D and 3D) (Section 5.4.5)
- Mitigating the problem of aliasing when correlating structural features using 2D seismic (i.e. absent 3D seismic) (Section 5.4.7)
- Direct detection of reservoir lithologies (Sections 5.4.3 and 5.4.4)
- Direct imaging of prospect scale features (Sections 5.4.2 and 5.4.6)

In all applications, the exploration impact is maximised when a variety of technologies are integrated intelligently. The relative contribution of any given technology to any specific exploration problem, is of course case specific.

The summary of the application for each case and the technologies that FTG are integrated with are shown in Table 2.5.

2.5.11 Marine Electromagnetic Methods

Electromagnetic methods measure resistivity (which varies over many orders of magnitude) and distinguish between lithologies of differing resistivities. Traditionally, such methods have been used for regional mapping and understanding, and to complement seismic methods in areas of complex geology. In recent times, EM methods (particularly CSEM) have reached most prominence in predicting the presence of absence of hydrocarbons. For a given reservoir lithology, replacing conductive pore fluids – salt-water – with resistive

Table 2.5 *Full tensor gradiometry and the exploration and production phase*

E&P phase	Basin screening	Access	Exploration	Prospect evaluation	Appraisal	Development	Production
Crustal seismic							▮
Gravity and magnetics	▮	▮	▮	▮			
Full tensor gravity	▮	▮	▮	▮			
CSEM							
Ocean bottom seismic							
Microseismic and passive seismic							
Conventional reflection seismic							

CSEM, controlled-source electromagnetic, E&P, exploration and production.
Green indicates a good fit; orange, partial fit; blank, not applicable.

hydrocarbon can result in a large increase in resistivity (often by an order of magnitude or more), and it is this change that is the target for many marine EM surveys; to distinguish between water and hydrocarbon bearing prospects.

Marine EM methods fall into two categories: natural source methods, which use EM fields generated in Earth's atmosphere and ionosphere as a source, and controlled source methods, in which the source fields are generated artificially by a source that is under the control of the survey geophysicist (Figure 6.3). Natural source methods (MT) may be applied for mapping large-scale background resistivity variations, depth to basement and features such as salt and/or basalt.

Marine MT is particularly useful for large-scale basin reconnaissance in areas where seismic methods perform poorly. In areas where potentially prospective sediments are obscured by high-velocity or seismically heterogeneous layers, the resistivity information derived from MT analysis can be used to define the geometry of such features: It is usually the case that bodies with high seismic velocity and impedance contrasts, for example, basalt or salt, also have a higher electrical resistivity than surrounding sediments. This information can then be used to construct more accurate velocity models, leading to improved seismic. Regional MT surveying of this sort can provide a useful precursor to more detailed seismic or controlled source studies.

Marine controlled source EM surveys are also used to identify structures such as basalt or salt as an aid to seismic mapping. Importantly, CSEM methods can also be used to identify potential reservoirs. Such methods are used to assist in prospect ranking, for which seismic data and well data are likely to be available to calibrate the EM response. To go further, reservoir appraisal is a still better constrained application for CSEM, that seeks to address the question of the extent of a known hydrocarbon accumulation away from well control, given knowledge of the seismic structure and properties.

The complexity in the application of the methodology is in being able to identify a potential reservoir horizon and then determine its lateral resistivity discontinuity as an indicator of hydrocarbons. The challenge is to discern this lateral change from the natural variability in the surrounding lithologies. A measurement of high resistivity does not guarantee the presence of hydrocarbons, commercial or otherwise. Many geological features are resistive. For example, tight sands, carbonates, volcanics or salt stringers may all have a resistivity signature that is identical to that of a hydrocarbon-saturated sand. As with any other geophysical measurement, resistivity must be interpreted carefully, alongside seismic or other information, within a calibrated rock physics framework in order to determine the underlying rock and fluid properties.

Thus in common with all potential field techniques, you have to have a conceptual model of what you are looking for before selecting the technology and its acquisition parameters. This model will have been constructed from other geophysical measurements, or in the case of onshore field work.

In specific circumstances, CSEM is invaluable and a low-cost way of constraining the subsurface interpretation. The case studies in this chapter provide excellent examples.

The first study (Chapter 6, Section 6.8) describes the application to improve improved sub-basalt imaging through combining marine CSEM and seismic data. There are a number of situations in which seismic struggles to provide an accurate representation of sub-surface structure, for example areas where potentially prospective layers are obscured by basalt or salt. In the first example, the benefits of combining EM and seismic data when imaging potentially prospective sub-basalt sediments is considered.

The second case study (Section 6.9) shows how CSEM is used to improve the reservoir characterisation through integrated interpretation of CSEM and seismic data. This involves identifying subsurface lithology and fluid properties using seismic and CSEM data.

The application of CSEM to the exploration workflow is show in Table 2.6.

2.5.12 Microseismic Technology

Microseismic technology as applied to the petroleum exploration and production industry finds its roots in classical seismology. It is now used to manage and track fractures in unconventional oil and gas extraction through 'fracking'. It is this frac monitoring application which is addressed in this chapter.

Acquiring microseismic data is no different than the acquisition of conventional seismic data with the exception that it is a passive endeavour that requires no sound sources like vibrators, airguns or dynamite. It does require the deployment of listening devices, geophones typically, and recording equipment of an appropriate bandwidth and sensitivity.

The purpose of acquiring microseismic is to answer questions such as

- What is the local direction of maximum horizontal stress as evidenced by the average azimuth of the hypocentre trend?
- What frac length was achieved?
- What frac height was achieved?
- What stimulated reservoir volume was achieved?
- Did the treatment stay withing the target reservoir zone?
- Did the treatment encroach on adjacent wells?
- What should the well spacing be?
- Did the treatment from adjacent stages overlap?

Table 2.6 *CSEM and the exploration and production phase*

E&P Phase	Basin screening	Access	Exploration	Prospect evaluation	Appraisal	Development	Production
Crustal seismic							
Gravity and magnetics							
Full tensor gravity							
Marine CSEM and MT							
Ocean bottom seismic							
Microseismic and passive seismic							
Conventional reflection seismic							

CSEM, controlled-source electromagnetic, E&P, exploration and production; MT, magnetotelluric. Green indicates a good fit; orange, partial fit; blank, not applicable.

- What should the stage spacing be?
- Were there changes in reservoir response to treatment along the well?
- If the treatment parameters were varied, how did the microseismic data respond to those differences?
- If diverters were deployed, were they effective?
- Did the frac excite movement on regional faults?
- Was there any evidence of casing failure, bad cementing of the well, plug failure or sleeves not performing as planned?

These issues are focused on understanding the fracture network. Microseismic can also assist in managing the production and the economic exploitation of a given rock volume.

- How is the proppant distributed within the stimulated reservoir volume?
- Can the microseismic data give insight on production rate and the estimated ultimate recovery (EUR)?
- Which parts of the reservoir or which treatment program gives rise to greater hydrocarbon recovery?
- What will be the drainage volume of the treated wells over time?
- What is the optimum development scenario?
- What is the stress distribution near the wellbore as it relates to both treatment, production and wellbore stability?

Thus in addition to locating the faults in space and time the technique is able to estimate the nature of the failure that created the observed signals, the focal mechanisms of the events.

The application of microseismic to the exploration workflow is show in Table 2.7.

2.5.13 Ocean Bottom Marine Seismic Method

Oil companies are under increasing pressure to maximise the recovery of hydrocarbons from existing fields, in production, rather than seek new resources through exploration. Effective reservoir management is predicated on an accurate description of the reservoir, its porosity/permeability distribution in three dimensions, and predicting the behaviour of the fluids in the reservoir, as oil is extracted. Clearly the behaviour of the fluids is dependent on the porosity/permeability distribution. Thus the dynamic behaviour of the fluids informs the static model of the reservoir. Understanding where remaining hydrocarbons are trapped in the reservoir allows appropriate interventions to take place, including drilling more production/injector wells, or changing the pressure regime in the reservoir through changing injector/production rates.

Table 2.7 *Microseismic technology and the exploration and production phase*

E&P phase	Basin screening	Access	Exploration	Prospect evaluation	Appraisal	Development	Production
Crustal seismic							
Gravity and magnetics							
Full tensor gravity							
CSEM							
Ocean bottom seismic							
Microseismic and passive seismic						■	■	■
Conventional reflection seismic	■	■	■	■	■	■	■

CSEM, controlled-source electromagnetic, E&P, exploration and production.
Green indicates a good fit; orange, partial fit; blank, not applicable.

Imaging both the reservoir and the fluids is fundamental to reservoir management. This is where ocean bottom nodes are becoming increasingly effective, because they produce the best possible images. In addition, they allow identical reflection seismic surveys to be acquired at different times (4D surveys). The reflection seismic data from the different surveys can be compared. Reservoir management changes fluid distribution and also creates geomechanical changes. 4D ocean bottom surveys deliver the highest quality seismic images, which makes such surveys most effective in reservoir management processes. Traditionally to improve seismic resolution, broadband towed streamer 3D has been the most common method for acquiring 3D seismic data and also for repeat surveys for reservoir monitoring purposes (4D). The unit cost varies between around $8,000 and $30,000 per square kilometre depending very strongly on the size of the survey area, local problem issues and obstructions, and of course on the state of the market. Repeat 'time-lapse' surveys, where (ideally) the tidal conditions during recording have to be carefully matched to an earlier survey, are at the higher end of the price scale.

Several techniques, referred to earlier as 'broadband', can be applied to towed streamer surveys which remove or reduce some of these limitations, albeit with substantial increases in cost. These surveys start from around $15,000 per square kilometre and can easily be five times the cost of conventional towed cable 3D.

At the top end of the ladder, placing detectors on the seabed allows most, if not all, of the technical limitations of towed cable work to be overcome. A huge inventory of results is now available, most of which are compellingly good.

Significant recent advances in technology have led to major reductions in cost which means more and more surveys are being acquired, which brings further reductions in cost. Although it remains more expensive than many towed-cable seismic configurations, it is difficult to challenge its positive economics. Even for a small but typical recent find containing say 25 mmbbl, the cost of the high-quality seismic data needed for exploration and development will be less than $1/bbl. 'Conventional' towed cable seismic data although cheaper, may neither find the reservoir nor contribute usefully to its development.

It has geometrical flexibility so it allows unrestricted source/receiver geometry, and it can provide data close to or under platforms.

Table 2.8 shows where ocean bottom node surveys are best applied. Unlike other techniques, this technology can completely replace conventional 3D seismic. Although very expensive, consideration is being given to very large 'sparse node geometry' which might just be viable for exploration work in difficult basins.

Table 2.8 *Ocean bottom seismic and the exploration and production phase*

E&P phase	Basin screening	Access	Exploration	Prospect evaluative	Appraisal	Development	Production
Crustal seismic							
Gravity and magnetics							
Full tensor gravity							
CSEM							
Ocean bottom seismic			�	▮	▮	▮	▮
Microseismic and passive seismic							
Conventional reflection seismic							

CSEM, controlled-source electromagnetic, E&P, exploration and production.
Green indicates a good fit; orange, partial fit; blank, not applicable.

References

Fraser, A. J., 2010. A Regional Overview of the Exploration Potential of the Middle East: A Case Study in the Application of Play Fairway Risk Mapping Techniques. In *Petroleum Geology Conference Proceedings.*

Grant, S., Milton, N. and Thompson, M., 1996. *Play Fairway Analysis and Risk Mapping: An Example Using the Middle Jurassic Brent Group in the Northern North Sea.* Norwegian Petroleum Society Special Publications. Oslo, Norway: Norwegian Petroleum Society.

Hoversten, G. M., et al., 2013. CSEM & MMT Base Basalt Imaging. In *75th European Association of Geoscientists and Engineers Conference and Exhibition 2013 Incorporating SPE EUROPEC 2013: Changing Frontiers.*

IEA (International Energy Agency), 2019. *World Energy Outlook 2019.* Paris: International Energy Agency.

Levell, B., Argent, J., Doré, A. G. and Fraser, S., 2010. Passive margins: Overview. In *Petroleum Geology Conference Proceedings.*

Nemčok, M., 2016. Models of source rock distribution, maturation, and expulsion in rift and passive margin settings. In *Rifts and Passive Margins*, 347–75. Cambridge: Cambridge University Press.

Roberts, D. and Bally, A., 2012. *Regional Geology and Tectonics: Phanerozoic Passive Margins, Cratonic Basins and Global Tectonic Maps.* Philadelphia: Elsevier.

United Nations, 2015. *Sustainable Development Goals.* New York: United Nations.

White, R. S., et al., 1987. Magmatism at rifted continental margins. *Nature*, 330, 439–44.

Wilson, H. A. M., Luheshi, M. N., Roberts, D. G. and MacMullin, R. A., 2010. Play Fairway analysis, offshore Nova Scotia. In *Proceedings of the Annual Offshore Technology Conference.*

Note

[1] An exploration play is defined as a specific combination of source, reservoir and seal. Most companies define a play by the stratigraphy and depositional environment of the reservoir.

3

Crustal Seismic Studies

JANNIS MAKRIS AND KEITH NUNN

3.1 Introduction

Studies of the earth's crust were initiated around 1910 by investigating the propagation of seismic energy generated by earthquakes. Mohorovičić was the first to define a discontinuity between the crust and the upper mantle from observations based on traveltime–distance plots of the 9 October 1909 earthquake that occurred in the Kulpa Valley in Croatia, at a 40 km epicentral distance from the seismological observatory. Mohorovičić and other earth scientists soon recognised that the so-called Moho discontinuity was a worldwide phenomenon separating lithologies of different physical properties, particularly compressional (P-wave) velocities and densities. They also realised that depth variations of this interface or transition zone were regionally significant and were dependent on the geological environment and history of a given area. That the crust was internally structured was recognised 15 years later by Conrad (Conrad, 1925, 1928) from earthquake observations. Jeffreys, in his book *The Earth* published in 1924, gave a summary of the structure of the earth's crust based on seismological findings and suggested that the earth's crust was composed of three layers: a 10 km thick upper layer with a P-wave velocity (V_P) of 5.4–5.6 km/s, an intermediate 20 km thick layer with a V_P of 6.2 to 6.3 km/s and a lower layer with a V_P of 7.8 km/s.

The nature of the Moho discontinuity was, and still is, in dispute among earth scientists. The general view, however, is that it corresponds to a rather abrupt change of the compressional wave velocity and density of the felsic crust from the mafic mantle. Efforts in mapping the structure of the earth's crust were subsequently improved by using controlled man-made explosions. They were initiated in the early 1920s by Angenheister (1927, 1928) and Wiechert (1926, 1929), who realised that by exploiting quarry blasts or large explosions from construction works, a more detailed and accurate definition of the earth's crust was possible. The few experiments performed in the 1923–35 period (see Byerly and Wilson, 1935)

39

were significantly intensified after World War II when systematic projects were organised. Since 1945 there have been many national and international projects aimed at mapping crustal structure and its physical and petrological properties. In the 1970s there was usually at least one large lithospheric profile recorded in Europe each year involving large numbers of recording stations provided and manned by groups of academic researchers. The 1974 Lithospheric Seismic Profile in Britain (LISPB) experiment was recorded along four lines totaling more than 1,000 km from Cape Wrath north of Scotland to Portland Bill in the English Channel using 5 deployments of 60 land recording stations at a spacing of 2–4 km with large explosive shots fired at 6 locations along, and at the ends of, the profile (Bamford et al., 1976). The variation of the Moho depth, the crustal structure and the upper lithosphere beneath the United Kingdom was published by Bamford et al. (1978) and has been used since as a reference and control point for several other seismic studies including some offshore surveys with possible hydrocarbon potential (e.g. Roberts et al., 1988).

An excellent worldwide review of the historical development of this deep crustal seismic research was published by Prodehl and Mooney (2012).

The present, short review of the importance of crustal studies and the techniques used to accomplish them will focus mainly on understanding the regional geology and those aspects significant for hydrocarbon exploration. Furthermore, the interaction of the seismic results with the physical properties derived from other geophysical fields such as density, susceptibility and conductivity will be discussed. Moreover, it will be stressed that combining geophysical observations with geological mapping is essential in selecting areas of economic interest and guiding exploration efforts.

The following sections will present the theoretical and experimental concepts that justify employing large-scale seismic experiments for delineating structures of interest in frontier plays or under geological conditions where conventional reflection seismic fails to produce conclusive results.

3.2 Some Theoretical Considerations for Studying Crustal Structure

Deep Seismic Soundings (DSS) or Wide Aperture Reflection and Refraction Seismics (WARRS) exploit wide-angle reflections and refracted diving waves, as well as normal incidence reflections in order to develop velocity-depth models of the crust and sediments. Seismic time sections from common shot gathers (CSGs) or common receiver gathers (CRGs) are used as input information for tomographic inversion and ray tracing forward modelling. Long-offset data (out to around 100 km or more) are required in order to observe the wave field of diving waves and wide-angle reflections that penetrate deep into the crustal section.

The modelling procedure was refined in recent years by seismic tomography (see Ditmar and Makris, 1996; Zelt and Barton, 1998), and the technology used has undergone significant development from the simple intercept time crossover formulas or the application of the Wiechert–Herglotz algorithm that produces 1D velocity–depth models to more sophisticated ray tracing techniques (e.g. Červený et al., 1977; Zelt and Smith, 1992). Ray tracing methods compute travel times and compare them with those observed. Finally, amplitudes are calculated for the observed branches of the CSGs or CRGs considering also the amplitudes and shapes of the wavelets, in order to compare the energy propagation along the various wave paths. Such algorithms were developed either using the reflectivity method or with finite difference solutions of the wave equation (e.g. Fuchs and Mueller, 1971). Fast solutions can be also obtained using the well-known Zoeppritz equations (Zoeppritz, 1919).

The importance of using wide-angle/aperture long-offset observations in mapping crust and sediments is demonstrated in Figure 3.1. The figure shows the energy conversion at the interface between two layers in a simple, two-layer model with the upper and lower layers having compressional P-wave (V_P) velocities of 4,000 m/s (4 km/s) and 6,500 m/s (6.5 km/s), respectively. The shear S-wave velocities (V_s) and densities (r_1 and r_2) of the two layers are shown in the figure. In the left side of the figure the black arrow is the incident ray of the P-wave. This is reflected and transmitted as P- and S-wave energy (blue and red arrows respectively). At subcritical angles (below $\sin^{-1}$ [4,000/6,500] $= 38°$ of incidence in this case), very little of the incident energy ($<9\%$) is reflected as P-wave and even less as S-wave energy.

Most of the P-wave energy at subcritical offsets is transmitted into the lower half-space as P-waves with a very small part as S-waves. However, at the critical

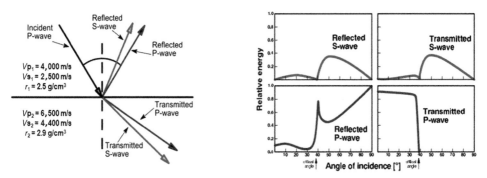

Figure 3.1 Reflection and transmission of a seismic ray interacting with the plain interface separating two isotropic half-spaces. The left side of the figure gives the velocities and densities used for the calculations. The right side shows the energy distribution of the P- and S-wave propagation.

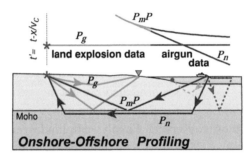

Figure 3.2 Seismic ray propagation paths for a simple two-layer case. Seismic phases indicated in the plot are Pg = P-wave propagating in the upper crust, PmP = P-wave reflected at the crust/mantle boundary (Moho) and Pn = P-wave propagating along the Moho in the upper mantle as a critically refracted ray path. In the upper part of the figure the 'reduced' travel time– distance plots of the three phases are shown with a 'reduction' velocity V_C = the velocity of the Pg phase such that this appears horizontal, with the faster Pn phase having a negative 'reduced' apparent velocity.

angle, >80% of the incident P-wave energy is reflected back into the upper layer and P-waves are not transmitted into the lower layer. Beyond the critical point most of the energy is reflected as P-waves with smaller amounts reflected and transmitted as S-waves.

The key point is that the wide-angle (supercritical) amplitudes of the reflected phases are much greater than the nearer normal incidence amplitudes used in conventional seismic reflection surveys (that use maximum source–receiver offsets of around the depth to the reflector) and it is these supercritical reflections that are mainly used in crustal seismic studies together with the refracted energy.

The lower half of Figure 3.2 shows a simplified, single-layer crustal model for an onshore/offshore survey with the three main crustal seismic phases viz Pg, the P-wave propagating in the crust as a continuously refracted 'diving wave'; PmP, the P-wave reflected at the crust/mantle boundary (the Moho); and Pn, the P-wave propagating along the Moho in the upper mantle. 'Reduced' travel time–distance plots of the same three phases are shown in the upper half of the figure with the 'reduction velocity' V_C = the Pg velocity such that the Pg phase appears horizontal and the faster Pn phase (usually around 8 km/s) appears with a negative 'reduced' apparent velocity. 'Reduced' travel time–distance plots are used in crustal seismic studies so that particularly crustal phases are more easily identified. Reduction velocities of 6 km/s are typically chosen (as in Figure 3.2) so that the direct crustal diving waves with apparent velocities close to 6 km/s appear near horizontal, and the refracted upper mantle Pn phase, which has a velocity around 8 km/s, appears with a negative 'reduced' apparent velocity. In very long offset surveys 'reduction' velocities of 8 km/s are also used to better display the upper mantle phases with the Pn phase appearing near horizontal.

Many WARRS studies also now use directly observed S-waves in onshore operations or converted PS-waves that can be detected on the geophone components of the recording stations placed at the sea bottom in marine operations (see Section 3.3) to produce V_s and hence V_P/V_s models to better constrain the lithology of the various crustal layers, as discussed in Section 3.4.

Figure 3.1 shows the key layer properties that determine the energy partitioning at interfaces between subsurface layers viz. the P-wave velocity V_P, the shear wave velocity V_s and the density ρ (shown as r in the figure). At normal incidence the P-wave amplitude reflection coefficient R just depends on the difference between the acoustic impedance (AI) $A = \rho \cdot V_P$ of the two layers and is given by the well-known equation

$$R = \Delta A / 2A, \text{ where}$$

$$\Delta A = A_2 - A_1, \text{ and}$$

$$A = (A_1 + A_2)/2,$$

with A_1 and A_2 being the acoustic impedances of the upper and lower layers respectively.

For the V_P and ρ values in Figure 3.1, $R = 0.3$ (and the relative energy coefficient is $R^2 = 0.09$ as shown in the figure). In practice the values of ΔA and hence R are much smaller across most subsurface interfaces (apart from the water bottom) and values of R are generally <0.1.

Both velocities (and densities) generally increase with depth (as compaction decreases porosity and the values tend to the matrix velocity and density as the porosity decreases to zero). They also depend on lithology (matrix) and geological age. Values of the various sedimentary rocks thus vary considerably with depth (actually their maximum depth of burial), and geological age and P-wave velocities are reasonably well predicted by the well-known Faust relationship (Faust,1951):
$$V = 1.45\,(ZT)^{1/6},$$
with V in km/s, depth Z in km and geological age T in millions of years.

This relationship would predict a value of 2.8 km/s for a mid Eocene (45 my) sedimentary layer at a depth of 1.5 km and a value of 3.8 km/s for a mid Cretaceous (107 my) layer at a depth of 3 km.

Average values for the commonly occurring sedimentary rock types, together with average values of the upper and lower crustal layers and the upper mantle, are provided in Table 3.1.

In crustal studies velocities are usually given in km/s ($= 10^{-3}$ m/s), as already used in this chapter. In the table, densities are given, as in Chapter 4, in

Table 3.1 *Average values of sedimentary and crustal rock velocities, densities and AIs*

	V_P (km/s)	V_S (km/s)	ρ (g/cm^3)	AI (km/s·g/cm^3)
Water	1.5	-	1.0	1.5
Unconsolidated sediments	1.8	0.7	2.0	3.6
Sandstone	3.2	1.6	2.3	7.4
Shale	3.4	1.5	2.4	8.2
Salt	4.5	2.6	2.2	9.9
Limestone	5.4	2.7	2.5	13.5
Dolomite	5.8	3.0	2.8	16.2
Granite	6.0	3.45	2.65	15.9
Basalt	6.4	3.55	2.8	17.9
Upper crust	6.0	3.45	2.65	15.9
Lower crust	7.6	4.4	2.8	21.3
Upper mantle	8.0	4.6	3.2	25.6

g/cm^3 ($= 10^{-3}$ kg/m^3) and Acoustic Impedances in km/s · g/cm^3 ($= 10^{-6}$ kg/s/m^2). For example, for granite the P-wave velocity is 6 km/s $=$ 6,000 m/s (in SI units), the density is 2.67 g/cm^3 $=$ 2,670 kg/m^3 (in SI units) and the acoustic impedance is 16.0 km/s · g/cm^3 $=$ 16 $\times$ 10^6 kg/s/m^2 (in SI units).

For most consolidated sediments (with the notable exception of salt) and many igneous rocks there is a good relationship between V_P and ρ, viz. the well-known Gardner equation (Gardner et al., 1974):

$$\rho = 1.74V_P^{0.25},$$

where ρ is in g/cm^3 and V_P is in km/s (see Chapter 4).

The relationship between V_P and V_S in terms of the elastic constant Poisson's ratio σ is

$$V_P/V_S = \sqrt{(2 - 2\sigma)/(1 - 2\sigma)}, \text{ with } 0.5 \geq \sigma \geq 0.0.$$

For most consolidated sediments, σ varies between 0.1 and 0.4, with V_P/V_S varying between 1.5 and 2.5 with an average value of 2.0 (corresponding to a Poisson's ratio of 1/3). For crystalline rocks and the upper mantle, the average value of V_P/V_S is $\sqrt{3} = 1.732$ (corresponding to a Poisson's ratio of 0.25). There are significant lithological variations from these average values for sedimentary and crustal layers and being able to measure both V_P and V_S in crustal seismic studies enables estimates to be made of V_P/V_S for the various layers which can be used to assess the lithology of the layer where there can be ambiguity when using V_P values alone.

Given that velocities generally increase with depth, within a layer the velocity will increase with a small but positive velocity gradient, dV/dz (typically <0.1/s), which has the effect of causing the transmitted waves within the layer to continuously refract away from the normal as 'diving waves' which can return back within the layer (if the thickness is great enough) to the surface as a 'first arrival'. Given that V_P is always $> V_s$, 'first arrivals' are always P-waves and, given that velocity generally increases with depth, these first arrivals at increasing offsets come from increasing depths enabling crustal seismic studies to provide velocity information on the whole crustal section, as explained in the text that follows.

The layer velocities and layer thicknesses determine the travel times of the various reflected and refracted phases that return to the surface for a given source–receiver pair and it is the inversion of the large number of travel times recorded in a seismic survey that enables a velocity depth model $V(x, z)$ beneath the survey line to be established.

Developing a laterally varying velocity–depth model $V(x, z)$ is a top-down process based on tomography and forward kinematic and dynamic ray tracing. In Figure 3.3 the steps followed in these calculations are presented schematically. The 'first break' or 'gridded' tomography exploits only the 'first break' P-wave phases that arrive first on the records. Observations of P-waves at longer offsets, having penetrated deeper within the crust as discussed earlier, delineate higher velocities as depth is increasing. The P-wave velocity model is developed iteratively by a least squares approximation procedure that automatically fits the synthetically computed travel times to those observed. The velocity model is presented by vertical and lateral velocity gradients in the (x–z) plane that allow velocity perturbations necessary to optimise the fit between the computed and observed travel times. The first break picks are extracted from the CRGs or CSGs. The root mean square (RMS) values of the travel time differences between observed and calculated travel times are acceptable when they converge to the picking accuracy of the first breaks. This is usually in the order of 20–30 ms, depending on the noise level of the raw data. The initial model that constrains the space, in which the model is iteratively developed, is constructed from the apparent velocities of the

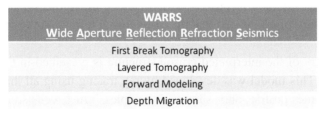

Figure 3.3 Computational steps followed in developing seismic velocity models from WARRS observations. This procedure can be applied for evaluating both P- and S-wave fields or PS converted phases.

CSGs or CRGs and estimates of the maximum thickness of the crust are calculated from wide angle crustal reflections (e.g. the PmP reflections from the Moho as shown in Figure 3.2).

The layered tomography that follows this initial phase uses the first break velocity model as a starting input. By assigning groups of reflection arrivals to assumed reflecting interfaces, defined at depth by abrupt changes of the velocity gradients, it is possible to develop iteratively the geometry of a reflecting interface. Its geometry and the lateral velocity variations in the layer above are automatically modified until the best fit between computed and observed travel times is achieved. The range within which the velocity perturbations can vary is constrained by the refracted (diving) waves and the velocities they have defined. This process follows a top-down evaluation layer by layer. The result obtained for each layer is then tested by forward modelling. Computed travel times have to fit those observed within picking accuracy and the synthetic amplitudes are obtained either using the Zoeppritz equations or finite difference techniques.

Once the P-wave velocity model has been established, the aforementioned procedure can be applied to develop the S-wave velocity field either from directly observed S-wave arrivals recorded by the horizontal components onshore, or, in marine cases, by PS converted phases (see the case studies in Section 3.4).

The common shot or common receiver gathers, which are the basis for the evaluation procedures described earlier, are prepared in SEGY formats. They can be processed by any commercially available software. Filtering and deconvolution are frequently applied to enhance the seismic signals and improve the signal-to-noise ratios. Static corrections are restricted to time corrections for instrumental drift only and do not consider topographic or bathymetric variations. In the modelling procedure described earlier we maintain the topography as it is. The reason is that seismic energy propagation is focused or de-focused by the shape of the earth's surface and synthetic amplitude calculations would be otherwise affected if the topography shape is artificially modified.

The upper part of Figure 3.4 shows an example of a CRG recording from the offshore East Africa crustal profile discussed in detail in the first case study in Section 3.4. The gather is plotted with a linear move-out computed using a reduction velocity of 6 km/s (so that the main crustal refraction phase appears near horizontal).

The geometry of the interpreted crustal structure is presented in the central part of Figure 3.4. This model was developed by ray tracing using all the 87 observed CSGs along the profile and shows a complex crust consisting of several sedimentary and crustal layers above the Moho discontinuity which shallows from a depth of 20 km at the western (continental) end of the section to a depth of 15 km at the western (oceanic) end. The synthetic travel times of the refracted and reflected arrivals computed from the ray tracing are colour coded and plotted over

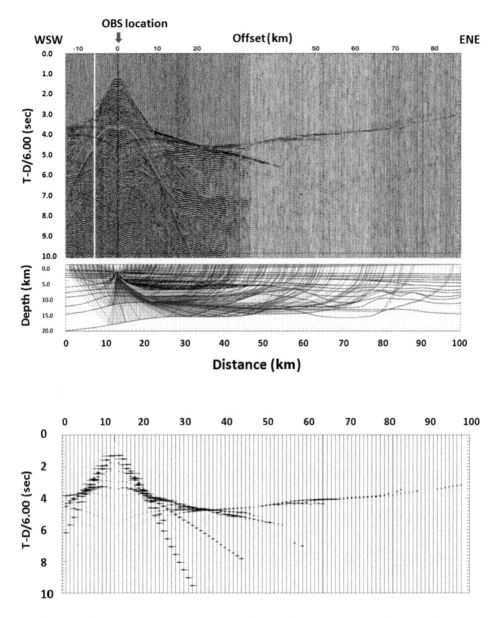

Figure 3.4 Example of a common receiver gather above its ray-traced seismic model and synthetic gather. The time section (upper part) is reduced with a linear move-out using a reduction velocity of 6 km/s and the model below was developed by ray tracing as discussed in the text (Makris et al., 2012).

the recorded data in the upper part of the figure. The lower part of the figure shows the synthetic amplitudes computed using the Červený code referred to earlier.

Figure 3.5 shows another CRG from the same offshore East Africa profile with the first arrival sedimentary phases Ps, the main crustal diving wave Pg and the

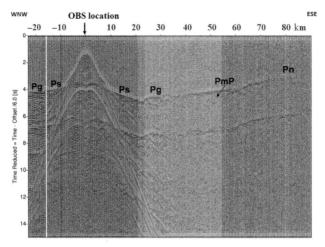

Figure 3.5 Example of a common receiver gather recorded offshore Kenya. The section is displayed with a reduction velocity of 6 km/s. Phases Pg, PmP and Pn are as in Figure 3.2. Ps denotes phases propagating through the sediments (Makris et al., 2012).

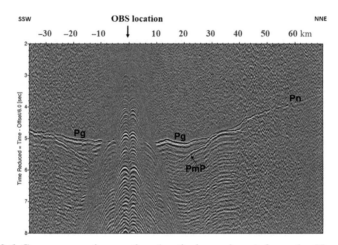

Figure 3.6 Common receiver gather (vertical geophone) from the Nova Scotia, North Atlantic passive margin (Makris et al., 2010). The section is displayed with a reduction velocity of 6 km/s. Shot spacing was 120 m and the P-wave phases Ps, Pg, Pn and PmP are as in Figure 3.5.

upper mantle Pn phase indicated together with the wide-angle Moho reflection PmP. Shots were fired every 100 m and energy propagated efficiently for more than 90 km offsets, which was adequate for delineating the crustal structure above the relatively shallow Moho (15–20 km as discussed earlier).

Figure 3.6 shows a further example of a CRG collected offshore Nova Scotia, Canada (Makris et al., 2010) with offsets out to 70 km. As in the offshore Kenya profile, the energy was generated using air gun shots and was recorded by autonomous, four-component ocean bottom stations with three geophone components and one hydrophone.

Both these examples are of ocean bottom seismograph data that were collected in studies of passive continental margins and were used to delineate the crustal structure, to define the continent–ocean boundary/transition and to develop geological models of rifting and continental break up as discussed further in the first two case studies in Section 3.4.

Onshore profiles using large explosions (fired at sea or on land) as sources are often much longer as necessary to study the Moho, which is at greater depths. The four lines of the 1974 LISPB experiment mentioned in Section 3.1 (Bamford et al., 1976) recorded data out to 600 km, and the Celebration DSS seismic experiment from Central Europe (Guterch et al., 2001) and the Trans European Suture Zone Polonaise'97 experiment (Grad et al., 2003) both used recordings out to 700 km.

Crustal Seismic Studies in both onshore and offshore cases produce data which are used to develop crustal models and to define the velocity structure in the uppermost part of the continental lithosphere. They are also used to map sedimentary basins in underexplored areas particularly where accessibility is difficult. The next section discusses the equipment used in onshore and offshore surveys.

3.3 Instruments for Collecting Crustal Seismic Data Onshore and Offshore

Any commercial system that can collect digital seismic data under field conditions, on- or offshore, is suitable for crustal and sedimentary basin studies. The minimum conditions they need to fulfil is to be autonomous, recording on batteries that last at least for 2–3 weeks. They need to have an internal quartz or atomic clock with high time stability and small linear drift. Recording GPS time is needed to calibrate the internal clock or to be directly used for onshore operations. A data digitiser and recorder with a large storage capacity of at least 8–32 GB and variable sampling rates (1–4 ms) is required. Data are usually recorded at a sampling rate of 2 ms. For land operations three recording channels for the one vertical and two horizontal component geophones are normally used. For marine operations four channel stations are needed for recording a hydrophone, as well as the three geophones. Stations have to be easy to program and download the recorded data without having to open the station. Batteries are usually rechargeable and are contained together with the electronic components and the communication system in a waterproof case that can withstand high pressure.

3.3.1 Onshore Applications

In land operations the three geophone components can be connected via cable to a central analogue to digital converter (ADC) board with CPU storing capacity. Geophones are usually spaced at 100–200 m and shots or vibrator sweeps are spaced at 500–1,000 m. CSGs at this spacing can sufficiently resolve the subsurface and permit depth migration of the recorded time field. Maximum shot-geophone offsets are variable and can extend from 50 to 150 km depending on the survey requirements. Much longer profiles are possible when large shots are fired in quarries, borehole patterns or under water. In such cases several tens to 100 kg explosives are simultaneously detonated. Seismic arrays must be continuously active, so that shooting can be optimised according to the operational requirements.

Several contractors and instrument manufacturers produce such systems. Their price and operational costs are very variable, and one has to consider the local survey conditions in order to decide whether to purchase or hire the correct system. Details of the different wireless and data transmitting systems are readily available on the Internet or from a consulting specialist. Figure 3.7 shows a selection of

Figure 3.7 Selection of land seismic stations suitable for operations under a wide range of field conditions.

9 different recorders which are suitable for land operations under a wide range of field conditions. All of them are stand-alone units operating with rechargeable batteries or solar panels. They are light weight and easy to transport in the field with cars or helicopters. They must be waterproof and operational under a wide range of temperatures. Handling must be simple, so that they can be deployed in large numbers for active or passive seismic projects. Types of sensors used can be geophones, broadband ultra-low-frequency seismometers or accelerometers.

3.3.2 Offshore Applications

Marine operations for crustal studies have a long history of academic development. Instruments for studying the world's oceans have been constructed in many countries (see e.g. Loncarevic, 1983) and ocean bottom seismographs (OBSs) have played an important role in understanding the nature and physical properties of the oceanic crust. This has had a decisive impact in constraining the petrological models and geological concepts of its development.

This technology gradually found its way from academic research into exploration for hydrocarbons. The changing oil industry needs and significant advancements in data storage capacity and efficiency of energy consumption opened new markets and the industry invested in new construction of autonomous seabed systems. Many of these are 'free falling' after release from the deployment vessel and their position on the seafloor is located by acoustic triangulation from air gun shots at the sea surface. Some more sophisticated systems are placed on the seafloor by remotely operated vehicles (ROVs), particularly if repeatability of positioning is required. The housing of the electronic components uses either titanium or aluminum cylinders or spheres made of glass, titanium or aluminum. Such systems are operated in widely spaced node geometries for crustal and regional studies or densely spaced for studying basins and structures in the sedimentary cover. The operational efficiency of such systems and the reduction of production costs have radically changed the experimental and survey possibilities. From a few units sparsely distributed, hundreds of nodes in 2D and thousands in 3D geometries and also for 4D applications are now being deployed.

Figure 3.8 shows a selection of OBSs currently in use for offshore crustal and basin studies. The differences in construction, operational requirements, cost of production and manpower needed to operate them are significant and the input and advice of specialists are generally required to select the best system for a specific application.

Free-falling instruments are significantly cheaper to operate than those having to be deployed by ROV's. The well-developed tomographic velocity modelling methods and advanced depth migration techniques make it unnecessary to place

Figure 3.8 A selection of ocean bottom seismographs used for crustal studies and hydrocarbon exploration projects.

ocean bottom nodes on exactly predefined positions for most applications. Acoustic positioning using the shots fired at the sea surface and the speed of sound in the sea can locate stations on the sea floor very accurately after they have been deployed by free fall. Free falling OBS operational costs are less than one third of the costs of ROV-operated systems usually producing velocity models as accurately as those obtained by ROV-deployed nodes.

Figure 3.9 shows one OBS developed by Geosyn Geophysics and used for both passive and active seismic studies. The electronic components of the OBS are placed in a 17- or 13-inch glass sphere, protected by a plastic housing. The OBS can operate in water depths up to 6,700 m or 9,000 m, depending on the thickness of the glass. The internal clock of the system is initially triggered by a GPS signal and has a time stability of 10^{-8} s. No special vessels are needed for deploying or recovering the OBSs. Programming the OBS station and data retrieving, as well as recharging batteries are performed without having to open the sphere. Communication between the OBS and the operations vessel is acoustically

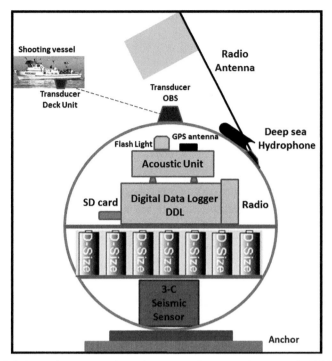

Figure 3.9 Schematic presentation of an ocean bottom seismograph (OBS) system developed by Geosyn Geophysics. All electronic components – the seismic sensors, the batteries, the analogue-to-digital converters, the radio for locating the OBS at the sea surface, the acoustic unit for communication between the OBS and the vessel, the flashlight for visual contact at night – are contained in the glass sphere. A deep sea hydrophone, a transducer for transmitting and receiving acoustic signals, a flag and radio antenna are placed on the external side of the sphere (Makris and Papoulia, 2017).

possible via a transducer system that is used for quality control and to activate the release mechanism. OBS stations are anchored on the seafloor by metal frames or sand bags.

Freely drifting buoys were widely used during the early stages of oceanic crustal studies but are not suitable for very long offset recording. Moored buoys have also been used but were never widely accepted due to mooring difficulties with increasing water depth.

3.4 Field Experiments and Examples of Crustal Seismic Studies

In this section four examples of crustal studies from different geological environments are presented. The use of such information for determining the physical properties of the crust is demonstrated. This information is used to constrain the geological concepts and delineate the structure of the sediments in basins with hydrocarbon potential.

3.4.1 The Passive Margin of Kenya, Indian Ocean

A crustal and basin seismic study of the passive continental margin of Kenya and its transition to the oceanic domain of the Indian Ocean was performed along a 180 km long WNW–ESE profile (Makris et al., 2012). The location of the profile is shown in Figure 3.10.

The water depths along the profile ranged from 1,750 to >3,750 m. Eighty OBSs were deployed with variable spacing along the profile. Shots were fired at 100 m intervals using air guns and the data were compiled in CRGs. An example of a CRG from this study is shown in Figure 3.5.

The aims of the survey were to study the crustal structure, to define the continent/ocean boundary (COB) and to collect information on the thickness and geometry of the sedimentary sequences. Identifying the COB transition offshore Kenya was the key target. This is essential for limiting the extent of the areas for

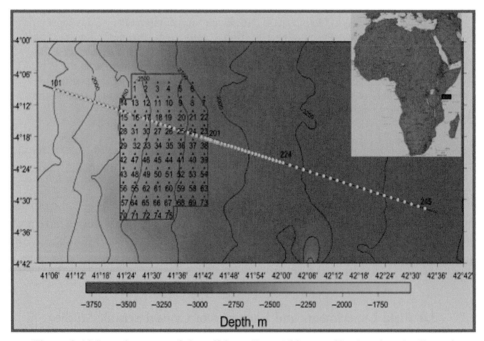

Figure 3.10 Location map of the offshore East Africa profile showing the line of ocean bottom seismograph stations and the bathymetry offshore Kenya (Makris et al., 2012).

hydrocarbon exploration. Offshore Kenya the main targets are located in basins that developed during rifting on the stretched continental crust.

As discussed in Section 3.2, first break tomography was first applied to produce an initial laterally varying P-wave velocity gradient depth model which reveals the velocity structure to about 14 km depth, as is shown in Figure 3.11. The lower crust is not delineated and it is difficult to locate the crust/ocean transition without using the complete wave field including the PS converted phases.

The first break tomography model that defines the velocity structure with velocity gradients was used as input for layer tomography and forward modelling by kinematic and dynamic ray tracing, using both the refracted and the reflected arrivals. This produced the complete section of the crust, from the continental to the oceanic domain shown in Figure 3.12. The continental crust is composed of two igneous layers with the stretched and rifted continental part underplated by a mantle intrusive as shown in Figure 3.12, which has a higher V_P velocity than the lower crust.

By also evaluating the observed PS converted phases, values of the V_P/V_S ratios could be computed for the key layers. Figure 3.13 shows the V_P/V_S ratios obtained

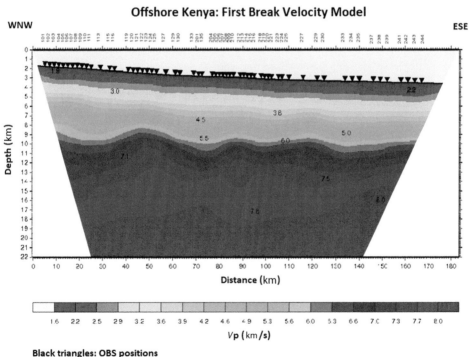

Figure 3.11 First break tomographic velocity model of the Kenya passive continental margin (Makris et al., 2012).

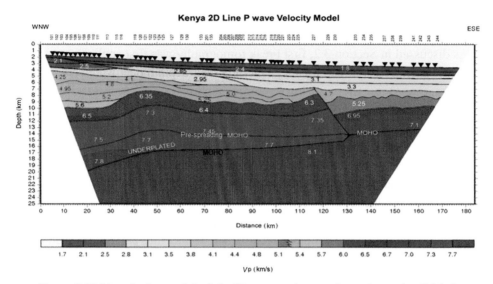

Figure 3.12 V_P velocity model of the Kenya passive continental margin. (Makris et al., 2012)

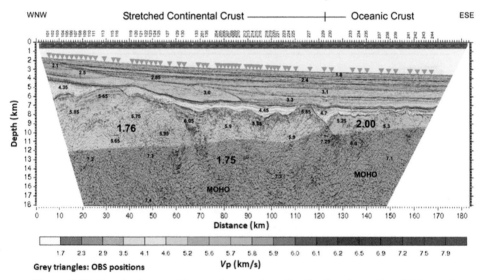

Figure 3.13 V_P velocity model superimposed on the depth migrated multichannel seismic data. V_P/V_S ratios are shown in the model. (Makris et al., 2012).

from the PS converted phases recorded by the horizontal components of the OBSs indicated in bold on the V_P velocity model. The procedure followed was to maintain the geometry as defined by the V_P evaluation and then, after identifying the correct PS conversion mechanism, develop the S-wave model. The validity of the S-wave model is tested by computing synthetic travel times and comparing them with those observed. The V_P/V_S ratio of 2.0 defines the oceanic crust, while the continental domain has V_P/V_S ratios of 1.75–1.76.

In this way it was possible to constrain the COB transition and also the nature of the underplated body in the continental domain, which is not serpentinised since its V_P/V_S ratio is 1.75. The true oceanic domain is located at a distance of 120 km of the model approximately 300 km from the Kenyan coast. It was identified by the change of the V_P velocity structure, the thickness of the layers and their V_P/V_S ratios.

Figure 3.13 shows the velocity model of Figure 3.12 developed from the OBS data superimposed on the depth migrated multichannel seismic reflection evaluation provided by ION-GXT offshore Kenya. The combination of the two data sets provided the optimum image for understanding the regional geology and tectonic deformation of the basin. The mapped lateral velocity variations in the basement provided additional information that helped the understanding of the regional geology and its influence in the development of the offshore basins.

The final model as seen in Figure 3.13 suggests that the stretched and rifted continental margin offshore Kenya provides interesting conditions for hydrocarbon exploration.

3.4.2 The Nova Scotia, North Atlantic Passive Continental Margin

A crustal and basin seismic study of the Nova Scotia shelf and slope and its transition to the oceanic domain of the West Atlantic was performed by recording a 400 km long OBS profile perpendicular to the Canadian coast overlying the eastern-most dip line (Line 2000) of the ION-GXT NovaSpan regional seismic reflection survey shown in Figure 3.14 (Wilson et al., 2010).

One hundred four-component ocean bottom seismographs were used with a variable spacing from 2.5 to 10 km, to address specific geological targets and optimise the field efforts. The denser OBS station spacing was centered on the expected position of the COB as indicated by the likely extension to the North-East of the East Coast Magnetic Anomaly (ECMA), the strong positive anomaly shown in red in Figure 3.14 which can be seen to decrease as it possibly deepens to the north-east.

Shots were fired by air guns at 120-m intervals and data were compiled in CRGs. An example of a vertical component CRG is shown in Figure 3.6. First break and layered tomography combined with two-point ray tracing were applied as discussed in Section 3.2 and, as in the previous case study in Section 3.4.1, both V_P and V_S models were produced. As mentioned earlier, the northern most half of the 400-km profile was located over line 2000, one of the multichannel seismic

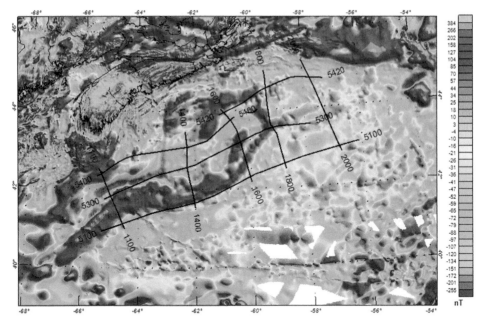

Figure 3.14 The North Atlantic Margin Offshore Nova Scotia showing the Magnetic Anomaly Map and the five dip lines and four strike lines of the ION-GXT NovaSpan regional 2D reflection survey. (Wilson et al., 2010).

(MCS) regional lines of the NovaSpan 2D reflection survey that had been shot by ION-GXT some years earlier using a 9 km streamer and shown in Figure 3.14. The more detailed sedimentary layer velocities from the conventional reflection data were integrated with the first 200 km of the OBS observations to aid the analysis.

The final velocity model is shown in Figure 3.15, which details the lateral changes of the crust over the 400 km long section from the continental domain, at the very shallow part of the shelf, to the oceanic domain at about 4,000 m water depth (Makris et al., 2012). The continental crust is 34 km thick and composed of 3 igneous parts beneath the very shallow shelf where it is covered by 2,500 m thick sediments. The sediments gradually thicken eastwards to about 5,000 m at the break of the shelf. Here the crust has thinned significantly and at 2,000 m water depth it is only 24 km thick. It also changes its structure, and from a 3-layered crust onshore it becomes a 2-layered continental crust beneath the deep water offshore. At the continent–ocean transition, at 250 km along the line, the crust is underplated by a mafic intrusion, which has a V_p/V_s ratio of 1.7 which implies that it is not serpentinised. Further evidence that the body was not a transitional, serpentinised layer was that strong reflections were observed from its base. Sediments are still fairly thick in the rifted zone and at 4,000 m water depth a 6,000 m thick cover overlies a significantly thinned continental crust. The oceanic domain is identified

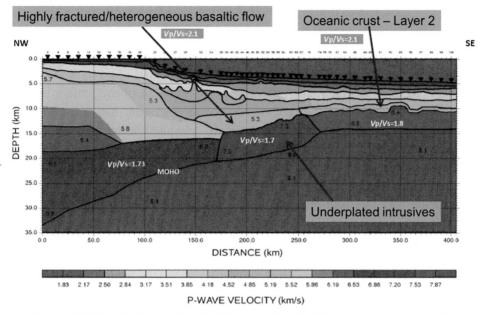

Figure 3.15 V_P velocity model of the Nova Scotia – N. Atlantic passive margin. V_P/V_S values are shown in the model. (Makris et al., 2010)

by a change in the V_P velocity, the crustal thickness, and the higher V_P/V_S ratios as seen in Figure 3.15.

The crustal seismic results were also considered together with the gravity data available along the line using velocity–density relationships for the various layers of the depth model to match the Bouguer anomaly along the line. The landward edge of the high velocity 'underplated body' shown in Figure 3.15 correlates with the East Coast Magnetic Anomaly (ECMA) shown in Figure 3.14 and explains the variation of the anomaly across the basin mentioned earlier. The deeper crustal velocities were also used as input to the reprocessing of the NovaSpan regional 2D data that was carried out soon after the crustal seismic study was completed as discussed further in Section 3.5.

These results were used both in a reconstruction of the pre-rifting stage of the Nova Scotia–Morocco margins and the rifting development of the Central Atlantic and impacted the Play Fairway analysis of the hydrocarbon potential of Nova Scotia (see Luheshi et al., 2012). That the geophysical evidence favours a volcanic rifting mechanism that extends across the Nova Scotian Basin provides good evidence for an Early Jurassic, restricted marine, source rock and the possibilities of a new oil play in the south-western half of the Nova Scotian margin as discussed further in Chapter 9 of this volume.

3.4.3 *A Crustal Profile of the East European Craton*

The third case study discussed is that of a continental crustal survey acquired across the East European Craton published by Bogdanova et al. (2006).

The southern margin of the East European Craton has been systematically studied using deep seismic sounding (DSS) crustal and lithospheric profiles during a decade of coordinated efforts by several East European and Scandinavian research groups. The synthesis of these efforts, presented in the overview paper published by Bogdanova et al. (2006), demonstrates in an impressive way how DSS studies can illuminate the geological picture of an area and constrain its development concepts.

In Figure 3.17 the crust and mantle structure obtained along the DSS profile (using data recorded from large explosions) shows a complicated three-layered igneous crust of 40–50 km thickness. A high-velocity mantle with a V_P of 8.5 km/s is mapped to 80 km depth and shows the deformation of the lithosphere. The seismic section between the Podolian Block in the south and the

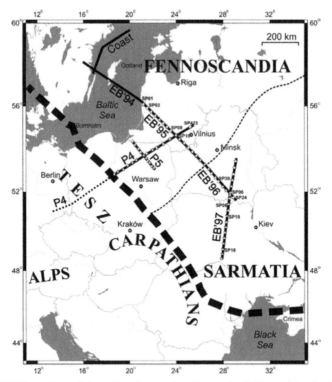

Figure 3.16 Location of the refraction and wide-angle DSS profiles in the south-western margin of the East European Craton, including the EUROBRIDGE'97 (EB'97) profile. (Bogdanova et al., 2006)

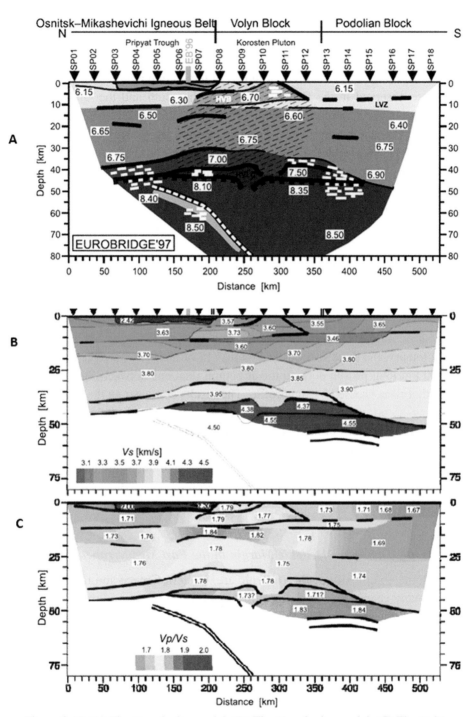

Figure 3.17 (A) The V_P velocity model. (B) The V_S velocity model. (C) The V_P/V_S ratio of a crustal cross section across the East European Craton. (Bogdanova et al., 2006)

Osnitsk–Mikashevichi Igneous Belt (OM-Belt) in the north is deformed and uplifted. The exhumed Korosten Pluton in the central part of the profile has a high V_P velocity value between 6.35 and 6.7 km/s and can be traced to a depth of at least 11 km. The V_P/V_S values in this central part of the section are systematically higher than those observed in the Podolian Block to the south and the OM-Belt in the north. This suggests that the Korosten plutonic block is affected by mafic plutonic intrusives from mantle-derived melts contaminated by the gabbroic lower crust. The very high values of the V_P/V_S ratios (around 2.0) in the upper part of the Pripyat Trough are associated with the sedimentary fill that was mapped to a 5 km depth.

As was also seen in the previous two case studies, it is important to measure both V_S and V_P velocity values and to study the V_P/V_S variations and their relation to the possible lithologies. Johnston and Christensen (1992) showed experimentally, and from V_S values obtained from crustal seismic experiments, that it is possible to relate the V_P/V_S ratios of igneous plutonic rocks to their composition. They also demonstrated that the V_P/V_S ratios are systematically different from continental to oceanic crustal domains. This can be exploited in order to define the continent ocean transition in rifted margins.

In summary, the three sections in Figure 3.17 of the V_P, V_S and V_P/V_S models of this 500 km long profile defined the structural relations between the three crustal blocks that were crossed by the profile. The results demonstrate the importance of measuring V_S as well as V_P velocities. The V_P/V_S models help in classifying lithological differences in the igneous crust and indicate processes that influenced their development. These experimental and evaluation procedures are not only valid for crustal studies but can be also applied for studying sedimentary basins for hydrocarbon exploration helping the interpreter to better understand the geological issues.

3.4.4 The Active Hellenic Margin in the East Mediterranean Sea

The tectonic complexity of the Hellenic arc, western Greece and the Ionian Sea, has for many years prevented successful exploration of the geological structures at depth in these regions. The Triassic evaporites that cover most areas of western Greece, both onshore and offshore, have restricted the depth of penetration of multichannel seismic reflection surveys. This has limited the exploration efforts for hydrocarbons to mapping secondary fields generated by hydrocarbon migration to shallow levels whereas the main structures of interest are sub-salt. As a result, many vital questions regarding the geometry and depth of the basins, crustal structure and its nature were poorly understood and were highly speculative.

In order to penetrate the sub-salt sediments and map the tectonic structure Makris and Papoulia (2014) used OBSs and stand-alone land stations along a series of wide aperture reflection/refraction profiles perpendicular to the coast and the main geological trends (see Figure 3.18).

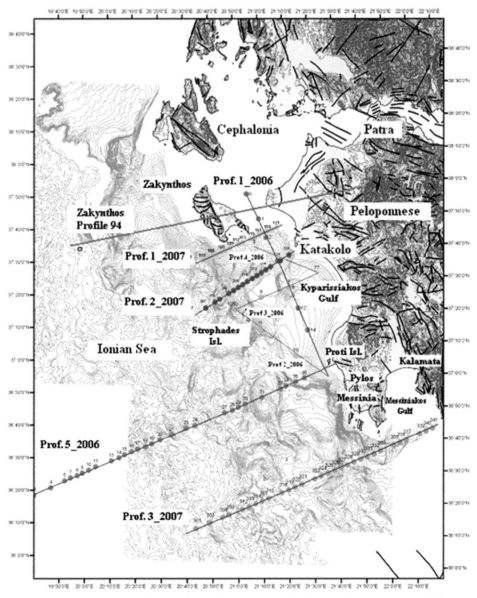

Figure 3.18 Location of the DSS profiles observed in the south-western part of the Hellenic arc. (Makris and Papoulia, 2014)

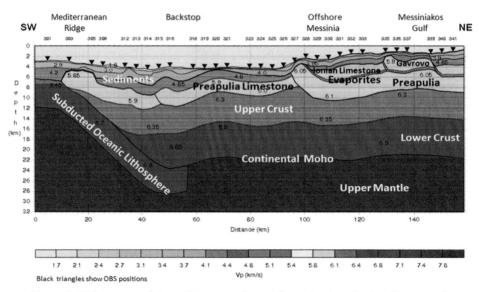

Figure 3.19 Structure of the sediments and crust from the deep Ionian Sea towards the Messiniakos Gulf (Peloponnese, Greece), derived from deep seismic soundings. (Makris and Papoulia, 2014)

Accurate velocity models were developed for the sediments and crust (Makris and Papoulia, 2014) and one of the profiles (Prof.3_2007, the southern-most near east-west profile in Figure 3.18) is shown in Figure 3.19. Offshore western Greece sediments in the basins are approximately 8 km thick. The Alpine high-velocity metamorphic limestones of the west Hellenic nappes, the Triassic evaporites and the Preapulia metamorphic limestones and sediments were mapped. The evaporites separate the upper from the lower limestones and 'lubricate' the westwards motion of the Hellenic nappes.

The thin evaporites indicated in Figure 3.19 were not directly mapped, as they were in other profiles further north although it was possible to identify reflections from the evaporite–Preapulia sediments contact. The thickness of the evaporite was estimated from travel time computations using the V_P velocity defined for the evaporite layer further north.

The limit of the continental crust offshore western Greece was located at a distance of 18 km along the profile, as shown in Figure 3.19. The structure and thickness of the sediments, as well as that of the continental crust offshore western Greece fulfil all the conditions for the generation and accumulation of hydrocarbons. Intense exploration activities have now started on/offshore western Greece after the Greek government lately defined concession blocks. The combination of multichannel reflection seismic with OBS data will be necessary in order to identify areas of economic interest and to map the sub-salt structures.

3.5 Integrating DSS/WARRS Velocity Models with Other Geophysical Information

Crustal seismic studies produce V_P, V_S and V_P/V_S information. These velocity models can be integrated with data from gravity, magnetic, magneto-telluric and heat flow observations. The distribution of density, susceptibility, apparent resistivity and the structure of the isotherms in the crust can be developed and used to constrain the geological models. This integrated approach is now widely used in regional geophysical and geological studies and it is also being applied in the exploration industry by, for example, Dell'Aversana (2014), who demonstrates the possibilities and advantages of simultaneously inverting different geophysical fields. If all these physical parameters are properly combined, one can produce multiparametric, robust geophysical models that are invaluable in understanding regional geological processes and also help the explorationist to define targets. Figure 3.20 shows a list of different geophysical methods and properties that can be constrained by the wide-angle seismic refraction and reflection results.

Integrated Geophysical Model		
Method	**Deliverables**	**Application**
Crustal Seismic	Vp, Vs velocity models Vp/Vs ratios	Regional geophysics & geology, basin studies, lithology, elasticity, anisotropy
Reflection Seismic	Deep reflectivity	Tectonic models, merging with Vp models into depth migration sections
Gravity	2D, 3D density models	Control of seismic results, isostacy
Magnetics	2D, 3D susceptibility models Definition of Curie surface	Mapping basin geometry, sea floor spreading models
Magnetotelluric	Definition of apparent resistivity in crust and upper mantle	Reliability of apparent resistivity increases significantly if constrained by the seismic model
Geothermics	Heat flow density maps	Computation of isotherms in the crust and mantle

Figure 3.20 Integration of crustal seismic velocities with other geophysical parameters.

3.5.1 Integrated Migration of Towed Cable Seismic Reflection and OBS Data

As discussed in the first two case studies in Section 3.4, integrating the long-offset seismic data from a marine OBS survey with the conventional towed cable seismic reflection data recorded over a coincident line can be of significant mutual benefit.

The shallow section of the OBS model can be much improved with the much finer sampling of the towed cable data (typically 25 m receiver spacing compared to the smallest 2.5 km OBS spacing in the Nova Scotia study discussed in Section 3.4.2). The imaging of the deeper section of the towed cable data can be improved with the better velocity control available from the much longer offsets recorded with the OBS survey (typically out to 100 km compared to the 8–12 km cable lengths used in long-offset 2D surveys; a cable length of 9 km was used in the Nova Scotia regional survey). The latter is of particular benefit when the imaging uses modern Pre-Stack Depth Migration (PSDM) algorithms. Modern Full Waveform Inversion (FWI) techniques have been one of the recent big advances in conventional seismic reflection imaging but the depth limitations are typically of the order of 30% of the cable length (8–12 km for most long-offset surveys as mentioned earlier): having much longer offset data (out to 100 km) offers the possibility of using FWI to provide optimum velocity models to the whole of the crustal section and much better PSDM sections.

3.6 Costs

The cost of a crustal seismic study clearly depends on the size of the area, the number of profiles and their total length, as well as the station density along the profiles. The number of available stations will determine the number of separate deployments and the duration of the field operations. Other factors include the location of the study, the type of equipment to be used, the personnel requirements and the mobilisation and demobilisation effort. Costs are similar to those of conventional 2D reflection surveys although, given the extra complexity of the specialised equipment, they are likely to be twice as expensive on a $/km comparative basis for both land and marine surveys.

For studies similar to the marine example discussed in Section 3.4.2 costs were typically around $4,000 per kilometer with a single 400-km line and OBS spacing of 2.5–10 km having an overall cost of up to $2 million. This cost includes all processing, modelling and interpretation expenses with these being less than 10% of the total financial cost.

Although crustal seismic information can be acquired also in a 3D mode with OBSs deployed along multiple lines and data acquired from multiple shot lines, in a similar way that ocean bottom node [OBN] surveys operate for exploration purposes (see Chapter 7 of this volume on OBN technology) there are hardly any field examples to demonstrate the advantages of collecting crustal seismic data in 3D. The high costs involved in operating the very large numbers of OBSs that would be required and the fact that there are no organisations that currently possess the number of stations needed have precluded such applications to date. In the future there could well be possibilities to organise such operations, particularly if efforts between academia and industry are combined.

3.7 Conclusions

Crustal seismic studies using active sources have played a vital role in understanding regional geology together with the concepts of plate tectonics. Experimental methods have developed over a period of more than 100 years from an academic level to one that satisfies the needs of the energy exploration industry. The tremendous developments in data storage and minimal energy consumption electronics have permitted the construction of stand-alone stations for land and marine operations that can be deployed and automatically operated using hundreds or thousands of stations. In parallel to the instrumental revolution, computer efficiency and the reduction of computational costs have enabled the development of techniques that permit fast processing of thousands of seismic traces and modelling methods that exploit the most complicated mathematical algorithms.

The refraction/wide angle reflection seismic method has long been the basis of global projects of international cooperation that study crustal properties both on and offshore in all possible environments. It has also demonstrated its fundamental value in the hydrocarbon exploration industry particularly in the early tectonic basin formation and evaluation stage of the exploration cycle and has also proved valuable in helping with several of the areas in the subsequent play fairway analysis stage.

The full value of merging and imaging long-offset conventional reflection seismic data with the velocities and data produced from refraction/wide-angle reflection seismic experiments is also now being realised, as is the value of integrating these velocities with other geophysical properties.

Overall crustal seismic studies provide a cost-effective approach that yields both regional models of high resolution as well as locally confined investigations and basin evaluations for new frontier areas.

Acknowledgements

We wish to thank Dr. Joanna Papoulia of the Hellenic Centre for Marine Research, Greece, for her considerable input in the preparation of this chapter. Anadarko of

Houston is thanked for the support in obtaining the Kenya refraction seismic data and permission to publish the data. ION-GXT provided the Kenyan and Nova Scotia reflection seismic lines: we acknowledge that their long-offset LithoSPAN 2D data have led to a significant improvement in the industry's ability to carry out regional, basin-wide geological and lithological studies. We thank the OETR Canada for supporting the Nova Scotia experiment and permission to publish the data. Finally, we thank the Greek Government and the European Commission for technical and financial support.

References

Angenheister, G. H., 1927. Beobachtungen bei Sprengungen. *Zeitschrift fuer Geophysik*, **3**, 28–33.

Angenheister, G. H., 1928. Seismik. In G. Geiger and K. Sheel(eds.), *Handbuch der Physik*, Vol. VI, 566–622. Berlin: Springer-Verlag.

Bamford, D., Faber, S., Jacob, B.., Kaminski, W., Nunn, K., Prodehl, C., Fuchs, K., King, R. and Wilmore, P., 1976. A lithospheric seismic profile in Britain. I. Preliminary results. *Geophysical Journal Royal Astronomical Society*, **44**, 145–60.

Bamford, D, Nunn, K., Prodehl, C. and Jacob, B., 1978. LISPB-IV. Crustal structure of northern Britain. *Geophysical Journal Royal Astronomical Society*, **54**, 43–60.

Bogdanova, S., Gorbatschev, R., Grad, M., Janik, T., Guterch, A., Kozlovskaya, E. and EUROBRIDGE and POLONAISE Working Groups, 2006. EUROBRIDGE: New insight in the geodynamic evolution of the East European Craton. Geological Society of London Memoirs. Boulder, CO: Geological Society of America. DOI: 10.144/GSL.MEM2006.032.01.36.

Byerley, P. and Wilson, T. J., 1935. The Richmond quarry blast of Aug. 16, 1934. *Bulletin of the Seismological Society of America*, **25**, 259–68.

Červený, V., Molotkov, I. A. and Pšenčik, I., 1977. *Ray Method in Seismology*. Univerzita Karlova, Praha.

Christensen, N., 1996. Poisson's ratio and crustal seismology. *Journal of Geophysical Research*, **101**(B2), 3139–56.

Conrad, V., 1925. Laufzeitkurven des Tauernbebens vom 28.11.1923. Mitt. Erb. Komm., Wien, Akad. Wiss., Neue Folge, no. 59.

Conrad, V., 1928. Das Schwadorfer Beben vom 8. Oktober 1927. *Gerlands Beiträge zur Geophysik*, **20**, 240–77.

Dell'Aversana, P., 2014. *Integrated Geophysical Models: Combining Rock Physics with Seismic, Electromagnetic and Gravity Data*. Houten, Utrecht, the Netherlands: EAGE Publications.

Ditmar, P. G. and Makris, J., 1996. Tomographic inversion of 2D WARP data based on Tikhonoc regularization. *Extended Abstracts, SEG Denver Annual Meeting*,

Faust, L. Y., 1951. Seismic velocity as a function of depth and geologic time. *Geophysics*, **16**, 192–206.

Fuchs, K. and Müller, G., 1971. Computation of synthetic seismograms with the reflectivity method and comparison with observations. *Geophysical Journal of the Royal Astronomical Society*, **23**, 417–33.

Gardner, G. H. F, Gardner, L. W. and Gregory, A. R., 1974. Formation velocity and density: The diagnostic basis for stratigraphic traps. *Geophysics,* **39**, 770–80.

Grad, M., Jensen, S. L., Keller, G. R., Guterch, A., Thybo, H. and Janik, T., 2003. Crustal structure of the trans-European suture zone region along POLONAISE'97 Seismic Profile P4. *Journal of Geophysical Research,* **108**(B11), 2541.

Guterch, A., Grad, M., Keller, G. R. and the CELEBRATION 2000 Organizing Committee, 2001. Seismologists celebrate the New Millenium with an experiment in central Europe. *EOS Transactions of the American Geophysical Union,* **82**, 529, 534–5.

Jeffrey, H., 1976. *The Earth: Its Origin, History and Physical Constitution,* 6th ed. Cambridge: Cambridge University Press.

Johnston, J. E. and Christensen, N. I., 1992. Shear wave reflectivity, anisotropies, Poisson's ratios, and densities of a southern Appalachian Paleozoic sedimentary sequence. *Tectonophysics,* **210**, 1–20.

Loncarevic, C., 1983. Instrumentation systems. In R. A. Geyer and J. R. Moore (eds.), *Handbook of Geophysical Exploration at Sea,* 317–32. Boca Raton, FL: CRC Press.

Luheshi, M., Roberts, D., Nunn, K., Makris, J., Colletta, B. and Wilson, H., 2012. The impact of conjugate margins analysis on play fairway evaluation: An analysis of the hydrocarbon potential of Nova Scotia. *First Break,* **30**, 61–72.

Makris, J., Nunn, K., Roberts, D. and Luheshi, M., 2010. A crust and basin study of the Nova Scotia Margin. In *Conjugate Margins Conference,* Lisbon, September 2010.

Makris, J. and Papoulia, J., 2014. The backstop between the Mediterranean Ridge and West Peloponnese, Greece: Its crust and tectonization. An active seismic experiment with ocean bottom seismogrpahs. *Bolletino di Geofisica Teorrica ed Applicada,* **55**(2), 249–79.

Makris, J. and Papoulia, J., 2017. The Geosyn Ocean Bottom Seismograph and its various applications for active and passive seismic observations. In *SEG Workhop OBN/OBS Technologies and Applications,* Beijing, China, Extended Abstracts. DOI: 10.1190/obnobc2017–24.

Makris, J., Papoulia, J., McPherson, S. and Warner, L., 2012. Mapping sediments and crust offshore Kenya, East Africa: A wide aperture reflection/refraction survey. In *SEG Las Vegas Annual Meeting,* Extended Abstracts. DOI: 10.1190/segam2012–0426.1.

Prodehl, C. and Mooney, W. D., 2012. *Exploring the Earth's Crust: History and Results of Controlled-Source Seismology.* Geological Society of America Memoir **208**. Boulder, CO: Geological Society of America.

Roberts, D. G., Ginzberg, A., Nunn, K. and McQuillin, R., 1998. The structure of the Rockall Trough from seismic refraction and wide-angle reflection measurements. *Nature,* **332**(14), 632–5.

Wiechert, E., 1926. Untersuchungen der Erdrinde mit Hilfe von Sprengungen. *Geologische Rundschau,* **17**, 339–46.

Wiechert, E., 1929. Seismische Beobachtungen bei Steinbruchsspregungen. *Zeitschrift fuer Geophysik,* **5**, 159–62.

Wilkens, R. G., Simmons, G. and Caruso, L., 1984. The ratio V_p/V_s as a discriminant of composition for siliceous limestones, *Geophysics,* **49**, 1850–60.

Wilson, H. A. M., Luheshi, M. N., Roberts, D. G. and MacMullin, R. A., 2010. Play Fairway analysis offshore Nova Scotia. In *Proceedings of the Annual Offshore Conference.*

Zelt, C. and Barton, P., 1998. Three dimensional seismic refraction tomography: A comparison of two methods applied to data from the Faeroe Basin. *Journal Geophysical Research*, **103**, 7187–210.

Zelt, C. and Smith, R. B., 1992. Seismic travel time inversion 2D crustal velocity structure, *Geophysical Journal International*, **198**, 16–34.

Zoeppritz, K., 1919. On reflection and transmission of seismic waves by surfaces of discontinuity, Nachrichten von der Königlichen Gesellschaft der Wissenschaften zu Göttingen, Mathematisch-physikalische Klasse, 66–84.

4

Gravity and Magnetics

ALAN B. REID

4.1 Introduction

Gravity and magnetic surveying have both played a significant role in petroleum prospecting over most of the history of the subject, have undergone evolutionary and revolutionary changes in that time and, in their modern forms, remain a valuable pair of tools in the explorer's toolbox. Modern instrumentation, navigation and computing power have made potential fields methods even more cost-effective than was historically the case.

Gravity and magnetism follow similar physical laws, but the detailed difference in those laws and the quite different distributions of density and magnetisation within the earth typically lead to quite different signatures and applications. In consequence, they often provide complementary information and also work very effectively in integration with electromagnetic and seismic surveying.

Table 4.1 summarises the broad properties of the various methods in exploration and Table 4.2 the geological scenarios in which they are generally applicable.

4.2 Units

This chapter employs International System (SI) standard units throughout, with a small number of subject-specific exceptions. The exceptions are shown in Table 4.3.

A detailed discussion of units is to be found in the text in the section headed 'Technical description'. Table 4.5 offers a useful range of unit conversions. Table 4.6 provides typical parameter values.

4.3 Historical Review

Gravity surveys have found application in petroleum prospecting since the early 1930s, following the discovery of the Nash Salt Dome, Texas, in 1924, using a

Table 4.1 *Broad properties of gravity, gravity gradient and magnetic data*

Application/ technology	Gravity	Gravity gradient	Magnetic
Depth sensitivity	Emphasises large, deep structure	Most sensitive to shallow structure (optimum at depths < 3 km)	Sensitive to magnetic rocks at all scales
Responds to	Lateral and vertical density changes	Lateral density changes	Lateral magnetisation changes
Costs/ relative	Moderate $60–70/km	Expensive (needs close spacing) $120–200/km	Low cost, <$20/km

Table 4.2 *Geological scenarios*

	Complementary data available	Gravity	Gravity gradient	Magnetics
Frontier	No wells, little or no seismic, topography, bathymetry, satellite and air photo imagery, satellite gravity	Basin delineation, structural controls	Cost-effective in special situations (salt structure, shallow faulting)	Regional structure and faulting, depth to basement, location of igneous intrusives and extrusives
Exploration	Some wells, sparse seismic	Fault delineation, salt delineation, integrated modelling	Detailed fault and salt delineation, especially at depths < 3 km	Fault connection, magnetic seep detection, using high-resolution aeromagnetic, detailed delineation of igneous interfering bodies
Appraisal	3D seismic and wells	Pre-stack depth migration (PSDM) velocity modelling, integrated modelling	Integrated modelling of challenging structures (e.g. salt)	Seldom used
Field development and monitoring	3D seismic, wells and production data	Precise density logging, production and water-flood monitoring with 4D gravity	Production monitoring with 4D gravity gradient	Location of old wells and pipelines

Table 4.3 *Industry versus SI units*

Parameter	Industry Unit	SI Unit
Density	g/cm^3	$kg \cdot m^{-3}$
Gravity anomaly	mGal, μGal	$m \cdot s^{-2}$
Seismic P-wave velocity	m/s and (sometimes) ft/s	$m \cdot s^{-1}$

torsion balance (a tedious but effective gravity gradiometer), and progressed rapidly with the development of portable gravimeters, most notably the LaCoste and Romberg zero length spring gravimeter in 1934. Modifications of that design went underwater, down boreholes, onto ships and eventually into the air (Hammer, 1983). Other instruments like the Askania and Bell marine gravimeters offered significant improvements. After a good start on salt problems, gravity surveys extended into basin delineation and tracking of major structural features such as faults and contacts. They were useful essentially any time there was a lateral change in rock density.

Many of the world's known basins (and much of the earth's surface) were covered with gravity observations at a variety of station densities. Such initiatives covered the whole of the former Soviet Union at a station spacing approaching 1 station per 4 km^2 (an amazing resource). The Ghawar (Saudi Arabia) and Burgan (Kuwait) super-giant fields were delineated with gravity data (Pawlowski 2020). In much of North Africa (e.g. the Sirte Basin, Libya, notably the Sarir Field), gravity was used to delineate structural highs, which hosted traps. Iraq was covered with a similarly high gravity station density, with amazing imaging of the basin architecture. The extraordinary diapirism of the Pricaspian Basin is spectacularly imaged by the gravity. Gravity thus became a very heavily used tool to 'zero in' on highly prospective areas. Gravity (often airborne) is being used today to delineate prospective basins in frontier areas, such as North-West and East Africa (e.g. Algeria, Mauritania, Mali, Uganda, D. R. Congo) and Australasia.

Gravity measurements have also found applications in monitoring production and (downhole) as a very sensitive density log.

Airborne survey magnetometers were originally developed during World War II for submarine chasing, with Gulf Oil's fluxgate system and the proton precession magnetometer leading. The primary early geophysical applications were identification of igneous intrusives and extrusives within sedimentary systems, the estimation of depth to crystalline basement and the delineation of structural features affecting both basement and overlying sedimentary features.

Much of the world was covered (at a variety of line spacings and survey heights) by aeromagnetic surveys, sometimes by systematic government-funded surveying, sometimes by petroleum and mineral exploration companies, and non-exclusive surveys and 'group shoots' by contractors. Systematic surveys cover all of Australia, Botswana, Canada, China, the Former Soviet Union, Finland, Guinea, Namibia, Sierra Leone, Tanzania, South Africa, the United Kingdom and Zimbabwe. This list is certainly incomplete.

4.4 Brief Resumé of Instrumental Development History

In the past few years, there have been significant improvements in what is possible, brought about by improved electronics, smarter sensors and better location, velocity and acceleration measurement using the best possible global positioning system (GPS) measurements. Airborne gravity measurement is extremely challenging, because aircraft vibration and turbulence can easily exceed 100,000 mGal, and we seek noise levels of 1 mGal or better. But such accelerations can now be measured precisely and, coupled with sufficiently linear sensors, can be removed from the observed data. Current instrumentation is listed in Table 4.4.

Going back to its roots, the other strand of gravity survey development has, of course, been gravity tensor gradiometry, which is the subject of Chapter 5 of this volume.

Table 4.4 *Recent gravity instrumental developments*

Application	Manufacturer/supplier	Model	Remarks
Land gravity	Scintrex/LaCoste	CG-6 Autograv	Zero-length quartz spring
	ZLS	Burris	Zero-length metal spring
Absolute land gravity	Scintrex/LaCoste	A10	Weight drop fall timing
Marine gravity	Scintrex/LaCoste	MGS-6	Zero-length metal spring
	ZLS	Dynamic Meter	with nulling feedback
	Sander Geophysics	Marine AIRGrav	Zero-length metal spring with vertical motion Triaxial accelerometers
Airborne gravity	Canadian Micro	GT2A sales	Triaxial accelerometers
	Gravity CMG	GT2A lease	Triaxial accelerometers
	Operations	AIRGrav	Triaxial accelerometers
	Sander Geophysics	TAGS-7	Zero-length metal spring
	Scintrex/LaCoste		with nulling feedback
Borehole gravity	Scintrex/LaCoste	Gravilog™	Zero-length quartz spring

Early magnetometer developments were followed by the Overhauser magnetometer, a sophisticated development of the proton precession concept, and the alkali vapour (K, Cs and Rb) magnetometers, which rely on physics at the atomic level. The alkali vapour instruments have been sensitive enough to make good vector gradiometers (e.g. the TriAx™ system), but recent developments with superconducting quantum interference device (SQUID) magnetometers have produced practical airborne magnetic tensor gradient systems. The magnetic gradient systems have mostly found application in mineral rather than petroleum exploration. The leading alkali vapour magnetometer manufacturers are presently Geometrics (G-824A), Scintrex (CS-3) and GEM Systems (GSMP-35). GEM Systems make the leading Overhauser magnetometer (GSM-19). IPHT of Jena, Germany produces the only known fly-able SQUID-based tensor magnetic gradiometer, but it is at present exclusively available to the Anglo-DeBeers group.

Satellite gravity and magnetic measurements have generated useful regional datasets.

Interesting instrumental developments 'in the pipeline' include

- Improved gravity gradiometers
- Ground-based, or fly-able or absolute gravimeters and gradiometers based on either free-fall measurements or 'cold atom' gravimeters
- Microelectromechanical system (MEMS) land-based and borehole gravimeters
- Wider availability of magnetic tensor gradient systems
- Further improvements in satellite measurements

4.5 Data Compilations and Global and Satellite Data Sets

4.5.1 Regional Compilations

A series of projects at the University of Leeds set out to assemble all available ground gravity data, continent by continent. Notable success was achieved for a large proportion of the earth's surface with the important exceptions of Australia (already done by Geoscience Australia; Minty, 2011), Canada – already done by the Geological Survey of Canada (GSC) – and the USA, Saudi Arabia and India (impossible to get agreement). Many major oil companies were subscribers to these databases and hold them in-house, but they may be accessed today via GETECH plc (https://getech.com/).

The International Gravimetric Bureau (BGI: http://bgi.omp.obs-mip.fr) carried out further compilations, concentrating on French Africa. Data contributions came from government entities and major oil and petroleum and mining companies.

A similar set of projects at the University of Leeds (sometimes in partnership with the International Training Centre, Delft, and Paterson, Grant and Watson,

Toronto) also compiled magnetic survey data worldwide, with much the same national exceptions. Again they may be accessed today via GETECH plc (https://getech.com) or from Paterson, Grant and Watson (www.pgw.ca).

The British Geological Survey (BGS) offers its own onshore and offshore gravity and magnetic data covering the whole of the UK at variable resolution (www.bgs.ac.uk/products/geophysics/). The BGS in collaboration with Bridgeporth run Geovault^{TM} (www.geo-vault.com), which offers licensable gravity and magnetic data deposited by a variety of petroleum exploration companies. Many of the major offshore seismic contractors hold significant archives of profile gravity collected along with marine seismic work, although this has dropped off to some extent with the advent of reliable airborne gravity surveys.

Some large contractors (notably CGG: https://geostore.cgg.com/geostore/) hold large databases of non-exclusive aeromagnetic data.

4.5.2 Global Compilations

Satellite Geodesy

In parallel with active instrumental surveying, academic geoid studies use satellite-borne radar to measure the geoid height (essentially the sea surface) and from it, deduce the gravity field offshore (as shown in Figure 4.1). Today those data sets are available everywhere offshore for wavelengths above about 20 km. They are freely available from University of California, San Diego (UCSD) by internet download from (http://topex.ucsd.edu/cgi-bin/get_data.cgi). A similar data set, possibly at lower noise levels on the continental shelves is available from the Danish National Space Institute, at https://bit.ly/367Lv4Y.

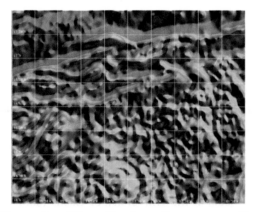

Figure 4.1 Satellite gravity offshore Honduras. Image covers 4° × 5°.

Both of these are extremely valuable free resources for regional offshore evaluation.

4.5.3 Global Gravity Models

Satellite (in-orbit) measurements have also generated lower resolution gravity data onshore, under the EGM-08, WGM2012, GRACE and GOCE programmes. These are best accessed via specialists because they can easily mislead the unwary.

Gravity Recovery and Climate Experiment Model

Gravity Recovery and Climate Experiment (GRACE) data were derived from a pair of satellites in low earth orbit following the same orbit, one close behind the other and measuring the distance between them to determine the in-orbit gravity gradient. From these data a global gravity model was deduced (Figure 4.2).

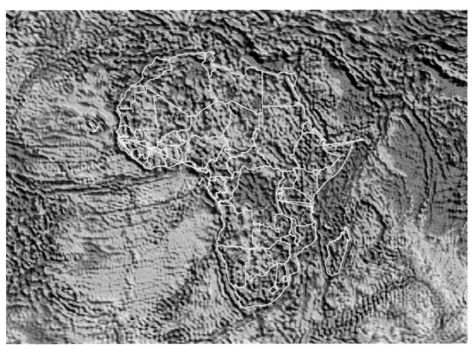

Figure 4.2 GRACE free air gravity image of Africa. Most of the prominent basins are imaged, including Congo Basin, many of the Algerian basins, the Murzuk, Sirte and Khufra Basins (Libya), South Sudan grabens, and the Luangwa valley (Zambia). The East African rift basins are barely resolved. The Karoo Basin gravity may be masked by its dense volcanic sequences. Major plate tectonic features offshore and crossing Africa are also visible. (Source: NASA)

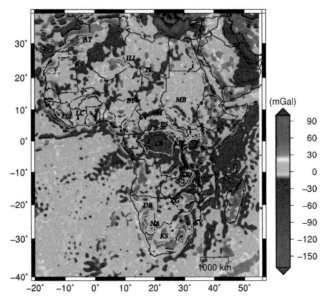

Figure 4.3 Free air gravity of Africa from GOCE observations. (Braitenberg, 2014)

Gravity Field and Steady-State Ocean Circulation Explorer Gravity Model

The European Space Agency's Gravity Field and Steady-State Ocean Circulation Explorer (GOCE) programme employed a satellite with an internal gravity gradiometer. The results may be extracted and presented in various ways, but the most useful for exploration may well be the free air gravity anomaly. GOCE seems to be showing better resolution than GRACE, by a factor of about two (Figure 4.3). It is possible to extract GOCE information for any given area from http://icgem.gfz-potsdam.de/.

Earth Gravity Model 2008

The Earth Gravity Model 2008 (EGM2008) and World Gravity Model 2012 (WGM2012) (https://bit.ly/3bGMLx1) (Bonvalot, 2012) combine satellite and ground gravity data (where available) but sometimes show implausibly fine onshore detail. This extra detail has been derived from a kind of 'reverse Bouguer correction' calculated from topography and is therefore not usually reliable as true gravity data. An example of real, data-supported gravity and spurious extra detail (from Guinea, West Africa) is shown in Figures 4.4 and 4.5.

4.5.4 Global Magnetic Model

Magnetic Compilation

EMAG2 is a world-wide magnetic data compilation available from the US National Oceanic and Atmospheric Administration (NOAA: https://data.noaa.gov/) free at

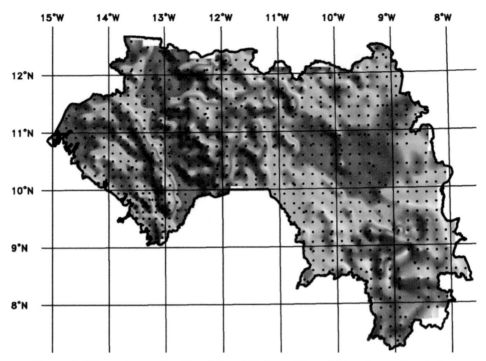

Figure 4.4 Free air gravity, Republic of Guinea, West Africa, from all existing ground gravity stations. (Data source: Bureau Gravimetrique International [BGI]). Gravity stations are plotted.

limited resolution. It is based on satellite measurements but includes smoothed airborne and ship-borne magnetic data. EMAG2 has a grid spacing of 2 arc minutes (~4 km). It is compiled at an effective elevation of 4 km and is useful for purposes of examining regional effects, as in the NW Pacific (Figure 4.6), where classic reversal-induced 'magnetic striping' is displayed in spectacular style, or, at a larger scale, in the S Atlantic (Figure 4.7).

EMAG2 is available from http://geomag.org/models/emag2.html.

4.6 Physical Properties

Modelling requires the use of sensibly chosen rock physical properties. Rock densities are seldom outside the range 2–3 g · cm^{-3}. The obvious exceptions are water (1.00–1.03 g · cm^{-3}, depending on salinity), and ore minerals, which can go as high as 5 g · cm^{-3}, but are seldom encountered in petroleum exploration. Rock magnetisation (typically specified via susceptibility) effectively varies from a small negative value (paramagnetic salt) to extremely high values greater than 1.0 SI units.

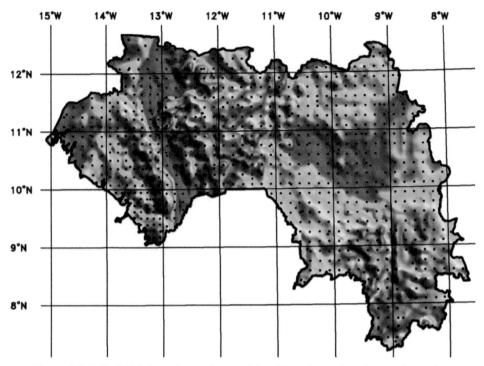

Figure 4.5 EGM2008 free air gravity model, with real gravity observation points overlaid. The extra apparent detail, unsupported by real data, comes from assumptions based on the topography. It is NOT derived from gravity observations and should not be used to draw exploration conclusions.

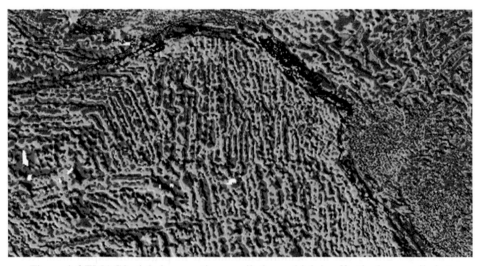

Figure 4.6 EMAG2 global magnetic compilation, which is effectively compiled at 4 km elevation. NW Pacific is shown. Source: NOAA.

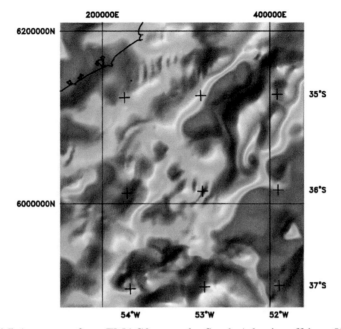

Figure 4.7 An extract from EMAG2, over the South Atlantic, offshore Uruguay. The magnetic anomaly (see Section 4.10.4 for magnetic nomenclature). Strong effects corresponding roughly to the continental margin are apparent. (Data source: NOAA)

Most specialist practitioners can be expected to have a working knowledge of likely density and magnetisation properties of typical rock types. Sometimes that is all that is available. But it is obviously preferable to use measured values where available. National Geological Survey organisations typically maintain such databases and will usually respond helpfully to specific requests but will sometimes require a fee. For gravity modelling over a basin, it is often possible to obtain access to one or more composite logs, which normally include lithology and density. Magnetic properties are logged in a usable form much less often. Sometimes it is possible to make measurements on core using a susceptibility meter or to commission measurement of remanent (that is, permanent) magnetisation. Measurement of remanence is the domain of specialist service companies and universities.

Some online rock physical property resources are provided by

- State of Victoria (Australia): http://earthresources.efirst.com.au/product.asp? pID=586&cID=42&c=93143
- University of Minnesota (Chandler and Lively, 2015): http://dx.doi.org/10 .13020/D63S3D)

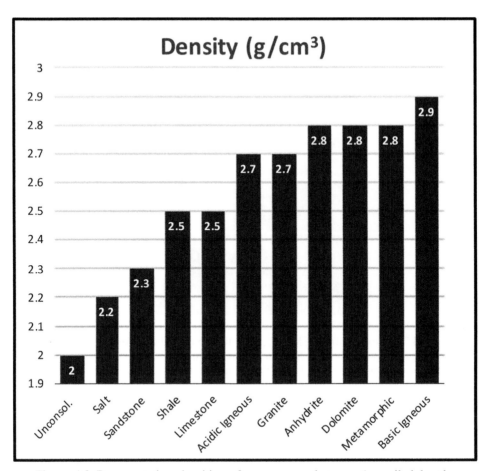

Figure 4.8 Representative densities of common rock types (compiled by the author).

4.6.1 Density

The typical variation in rock density with lithology is illustrated in Figure 4.8.

It is clear from Figure 4.8 that salt (halite) is anomalously low, at 2.2 g · cm^{-3}. This drives the whole phenomenon of diapirism and makes it easy to find salt diapirs using gravity (the earliest successful application of gravity to petroleum prospecting). Similarly, anhydrite has a very high density (2.8 g · cm^{-3}), giving rise to even more localised 'cap rock' highs within a salt diapir low. It is common practice to take the average density of crustal rocks to be 2.67 g · cm^{-3} (close to granite) for regional terrain correction purposes.

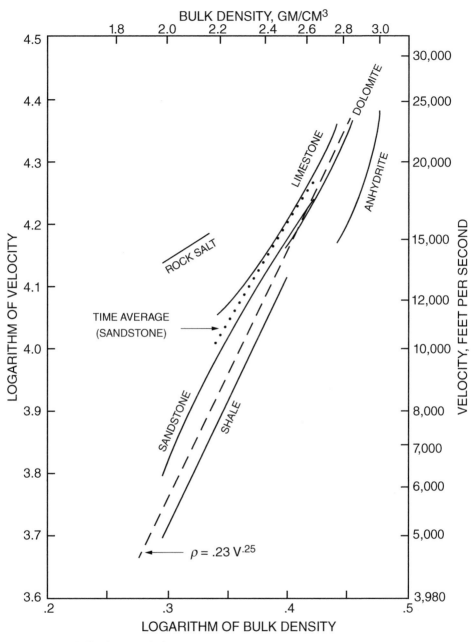

Figure 4.9 Gardner's relation.

Because gravity modelling is often done with the help of some existing seismic data, many workers find it convenient to use Gardner's relation (Figure 4.9; Gardner et al., 1974).

$$\rho = 1.741 \ V_P^{0.25},$$

Where ρ is density in $g \cdot cm^{-3}$, and V_P is P-wave velocity in km/s.

The relation becomes

$$\rho = 0.31 \ V_P^{0.25}$$

for P-wave velocities in m/s.

This relationship is purely empirical and is based on measurements in the Gulf of Mexico.

4.6.2 Magnetic Susceptibility and Remanence

Magnetic susceptibility expresses the degree to which any given rock mass can be reversibly magnetised in an ambient magnetic field (like the earth's). This is *Induced Magnetisation*, which will normally be aligned with the local geomagnetic field direction. The typical variation in rock magnetic susceptibility with lithology is illustrated by the diagram below (Figure 4.10) from EM GeoSci (https://em.geosci.xyz). The magnetic susceptibility is very roughly proportional to the volume % magnetite (Fe_3O_4), because magnetite is the only common

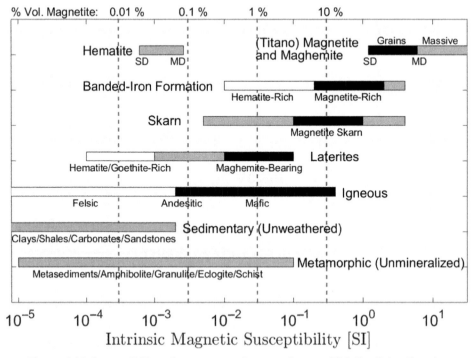

Figure 4.10 Susceptibility of common rock types. Source: EM GeoSci – Creative Commons Licence (https://creativecommons.org/licenses/by/4.0/)

magnetic mineral. More rarely, haematite (Fe_2O_3), pyrrhotite (Fe_7S_8), greigite (Fe_3S_4) and siderite ($FeCO_3$) may be present in sufficient quantities to be detectable using magnetic surveys. In sediments, pyrrhotite and greigite may be the residue of bacterial metabolism of petroleum in the presence of magnetite (Lovley et al., 1987), and may therefore be diagnostic of petroleum (Reynolds et al., 1990).

Acidic rocks and most arenaceous sediments tend to be weakly magnetic or effectively non-magnetic, whereas shales can be somewhat magnetic (possibly from high haematite content) and intermediate, basic and ultrabasic igneous rocks are increasing magnetic. The most magnetic rock commonly encountered in nature is banded iron formation (BIF). This would normally be encountered in petroleum prospecting as a basement rock.

In addition to *induced magnetisation*, a rock may display *remanent magnetisation* (or *remanence*), which is a permanent magnetisation arising by various mechanisms during the geological past. Remanence may be in any direction, depending on the origin and history of the rock mass concerned. It is most often displayed by igneous rock masses such as dykes, sills, lava flows and intrusions, but can also occur by low temperature chemical changes, like oxidation or bacterial metabolism. Remanence gives rise to odd magnetic anomaly patterns, uncharacteristic of the local magnetic inclination. An example is seen in the East Irish Sea Case Study (Section 4.11.4).

4.7 Other Significant Developments

4.7.1 Satellite Navigation (GPS, GLONASS, Galileo)

The companion instrumental development has been the flowering of satellite navigation. Gravity and magnetic surveys require large location, velocity or acceleration-based corrections, so a precise knowledge of location, velocity and acceleration reduces the uncertainty (which appears as noise) in the final product. The leading practitioners conduct Doppler processing of the SatNav carrier wave to determine velocity directly and therefore acceleration with enhanced precision. Airborne gravity surveying is possibly the most challenging satellite navigation application extant, to correct the effects of aircraft movement at the mGal level. In addition to the US-operated Global Positioning System (GPS), satellite navigation systems are increasingly taking advantage of the Russian GLObal NAvigation Satellite System (GLONASS) and now the more precise specification of the European GNSS Agency's Galileo system (all the way down to recent smartphones). This precise navigation also comes at vastly reduced cost, compared to traditional survey techniques, with consequent cost reduction and increased flexibility for ground-based surveys.

4.7.2 SRTM and Other Terrain and Bathymetry Data Sets

The recent free availability of good-quality land elevation data almost world-wide via the SRTM (1 arc-second ~30 m grid) has also made an amazing difference to the subject. It is available from https://earthexplorer.usgs.gov.

Where necessary it can be supplemented by local airborne LIght Detection And Ranging (LIDAR) surveys or commissioned work on satellite data. Similarly, free bathymetry data have come available in the form of the General Bathymetric Chart of the Oceans (GEBCO) www.gebco.net/. UCSD offers improvements on the bathymetry using satellite altimetry, but the improvements employ the satellite gravity data and in consequence there is circular reasoning, and the 'improved' data are **not** usable for bathymetry correction to gravity data.

4.7.3 Computing Power and Digital Signal Processing

Potential field surveys have benefited from digital signal processing and the extraordinary growth in computing capability in much the same way as seismic surveying has done, using many of the same concepts, most notably the fast Fourier transform (FFT). In consequence a line km of airborne magnetic data and a ground gravity station cost much what they did 30 years ago (in dollar terms), at significantly improved precision and accuracy.

4.8 State of the Art

From an instrumental viewpoint, current ground gravity surveys in representative petroleum exploration conditions deliver gravity values accurate to about 0.1 mGal. The best engineering-scale ground surveys achieve precisions of a few microgal. Good airborne gravity surveys will deliver mis-ties with a standard deviation of about 0.5 mGal for full wavelengths longer than about 7 km. The best airborne magnetic surveys have noise levels from all sources below 0.1 nT and are capable of delineating magnetic structure (faults) in sediments such as shales.

At the level of exploration applications, we see a strong move towards diversification and integration with other geoscience information sets. Some examples would be

- Mapping the continent-ocean boundary – essential in developing play concepts.
- Mapping the Curie point isotherm (see Glossary) – vital information for heat flow and hence petroleum systems modelling.
- Fault connection – We can map faults much more precisely by combining fault identification in section on sparse seismic lines with detailed mapping in plan using high-resolution aeromagnetic (HRAM) survey data for the fault connection.
- Bringing extra precision to magnetometer-guided directional drilling by providing precise local earth's field directions.

We are in the position that hardware, software and computing originally developed primarily for military purposes have had profound effects on our exploration technical capability. This is truly a case of 'beating swords into ploughshares' (Isaiah 2:4).

4.9 Technical Description

4.9.1 Scientific Basis of the Technology

Both gravity and magnetic fields are potential fields, so they follow similar mathematical laws. But the simplest gravity source is a small mass particle or 'monopole' whereas the simplest magnetic source is a 'small dipole'. Gravity 'lines of force' begin at infinity and end on masses. Magnetic 'lines of force' always form closed loops, so that there is always as much 'negative anomaly' as 'positive anomaly' associated with any collection of magnetised material. In addition, there are profound differences in the distributions of density versus the distributions of magnetisation among likely lithologies. In consequence, observed gravity fields look quite different from observed magnetic fields over the same geology and **yield independent geological information**.

4.9.2 Scaling

Gravity fields follow an inverse square law for the field fall-off with distance for a compact source. But the strength of the gravity anomaly from a source depends directly on its mass (volume × density contrast). Since volume varies as the cube of linear size, large structures have large gravity anomalies and gravity is best suited to delineating large structures (e.g. basins). It can be used at the civil engineering scale, but the anomalies are very small and great care is needed.

Magnetic fields fall off as the inverse cube of distance from a compact source, so the amplitudes of magnetic anomalies remain constant over varying scale. In consequence, magnetic survey data are equally good at delineating small-scale features (shallow faults, shallow lava flows), and large-scale features (basement intrusions, basin-bounding faults). One penalty is that widely spaced magnetic surveys are more likely to be misleading, because across-line undersampling can very easily yield grossly distorted maps. The phenomenon is known as 'aliasing' and arises whenever sampling (of anything) is done at a wider spacing than high-amplitude variations (Reid, 1980). Gravity surveys are less susceptible to this problem because the scaling laws have the effect of reducing the anomaly amplitude of small-scale structures.

4.9.3 Gravity Fields

The actual physics was fully developed in Newton's *Philosophiæ Naturalis Principia Mathematica* (1687). The advent of General Relativity has led to some modifications, but they're only important at astronomical scales (relativistic corrections are built into the GPS system, which is now so vital to exploration). With that single exception, we use Newtonian theory in petroleum exploration. You should be wary of any claimed new gravity technology which purports to be based on 'more modern concepts' than Newton's. That's the realm of voodoo geophysics.

In practice we exploit measured gravity variations (after significant corrections) to detect spatial or temporal variations in bulk rock density. No other rock property has any gravitational effect.

Measurement may be from satellites, on the ground, in ships, on the sea bottom, in aircraft or in boreholes.

Newton's law of universal gravitation states that

$$F = G M_1 M_2 / R^2.$$

The force F between two particles of mass M_1 and M_2 is proportional to the product of their masses divided by the square of the distance R between them. If the masses are specified in kilograms and the distance in metres, the force is in newtons and the gravitational constant

$$G = 6.672 \times 10^{-11} \text{ N} \cdot \text{m}^2 \cdot \text{kg}^{-2}.$$

We are interested in the gravitational effects of crustal rocks (the 'gravity anomaly'). But to isolate those effects we must remove effects from the earth as a whole, and to that end, we apply a series of standard corrections.

Gravity Field Units

There are no Imperial units in use in gravity work (except where feet and miles occasionally intrude). A gravity field is essentially an acceleration (the 'acceleration of gravity'), so its SI base unit is metres/second/second or $\text{m} \cdot \text{s}^{-2}$. The earth's gravity is often taken to be constant and have a value of $9.81 \text{ m} \cdot \text{s}^{-2}$. Geological effects are very small, so scientific work employs the gravity unit (or g.u.), which takes the value $10^{-6} \text{ m} \cdot \text{s}^{-2}$. Oil exploration tends to use the old cgs unit, based on centimetres rather than metres, and therefore uses the 'milligal' (mGal), or sometimes the 'microgal' (μGal).

$$1 \text{ mGal} = 10^{-5} \text{ m} \cdot \text{s}^{-2}, \text{ so } 1 \text{ mGal} = 10 \text{ g.u.}$$

The earth's gravity (g) is roughly 1 million mGal (981,000 mGal).

You will sometimes meet mixed units like gravity gradients expressed in mGal/mile or mGal/Kft, but these are best avoided unless you want to carry a lot of conversion tables.

The SI unit of density is $kg \cdot m^{-3}$. The density of water is $1,000 \ kg \cdot m^{-3}$. Since most users are more comfortable with the cgs unit ($g \cdot cm^{-3}$ or 'grams per cc'), the conversion factors between mGal and $m \cdot s^{-2}$ and between $kg \cdot m^{-3}$ and $g \cdot cm^{-3}$ need to be borne in mind in quantitative work.

Standard Gravity Corrections

Standard gravity corrections are a vital part of basic data processing. For the most part they deal with well-understood effects that are typically much bigger than the geological signal we seek to isolate. The acquisition contractor normally does these corrections. To be done properly, a gravity survey must be tied to a station that is part of the International Gravity Standard Network (IGSN). The IGSN is a network of standard stations, at which the absolute value of the gravity has already been established. The corrections include

1. Instrumental corrections (drift, calibration, absolute value). After these corrections we have the 'observed gravity'.
2. Subtraction of 'normal gravity'. This removes the earth's main gravity field, which has a very large constant part and a large latitude-dependent part (from the earth's rotation and flattening).
3. The 'Eötvös correction'. If the observations are from a moving platform (e.g. ship or aircraft) the platform velocity adds vectorially to the instantaneous part of the earth's rotation velocity and affects this correction very strongly. We therefore need to know the instantaneous speed and direction of motion quite precisely. Modern GPS-derived values are usually sufficient.
4. The free air correction. This corrects for the changes in gravity from differences in distance from the earth's centre. It appears as a vertical gravity gradient of value 3.086 mGal/m. For a precision of 0.1 mGal, we therefore need to know the observation height to a precision of better than 3 cm.
5. After the free air correction, we obtain the 'free air anomaly'. The process so far is essentially agreed and followed by all practitioners.

Bouguer and Terrain Corrections

If there is rock between the datum elevation and the station, and perhaps terrain surrounding the station, both will have gravitational effects. If we are modelling the gravity effects of the geology and matching the gravity measurements, we may find it convenient and conceptually simple to incorporate all this in the (3D) model. In that case, we would model the free air anomaly. If we are simply mapping and imaging the gravity and perhaps generating interpretive enhancements, we may not wish to be misled by topographic effects. In that case, we may perform a 'simple Bouguer' or 'Bouguer slab' correction and possibly also a 'terrain' or 'topographic' correction. The combined version is sometimes called a 'full Bouguer' correction. The Bouguer slab correction (in consistent SI units) is

$$g_{\text{Bouguer slab}} = 2\pi G \rho h$$

where G is the universal gravitational constant, ρ is the rock density and h is the slab thickness (generally taken to be the difference between the datum elevation and the station elevation). For gravity in mGal, thickness in metres and density in $g \cdot cm^{-3}$, the Bouguer slab correction is

$$g_{\text{Bouguer slab}} = 0.04192\,\rho h \text{ mGal.}$$

Since the Bouguer correction requires an assumed value of the local rock density, it treads into the realm of interpretation. Any Bouguer anomaly map MUST have its assumed Bouguer density and datum elevation included as 'key facts'. It is possible to model the Bouguer anomaly accurately by treating the Bouguer correction density as a 'background' density and inserting any model rock densities as differences ('contrasts') from that density. But the original correction density **must** be known.

Some workers have advocated the use of a variable Bouguer density, using plausible values from the observed surface geology. Although superficially attractive, it is all too likely that the 'variable density Bouguer' map will become separated from the density grid used and will be impossible to model later on. This approach is therefore not recommended.

Terrain corrections, which are an elaboration of the Bouguer correction, are important for precise work in difficult terrain. Formerly tedious graticule-based methods have been superseded by computer-based methods, but they require precise and detailed topography information near each gravity station and progressively less detailed information out to 167 km (the standard agreed for historical reasons). Airborne gravity and marine gravity surveys are usually easier to correct because the observation points are usually not very close to any strong topography changes.

Isostatic Correction and Residual

Much of the earth's crust is in isostatic equilibrium. Since crustal rocks are typically lower density than mantle rocks, this is another way of saying 'the crust floats on the mantle' and Archimedes' principle applies. That means in turn that mountains will have 'roots' and oceanic crust is thinner than continental crust. This

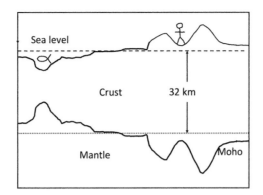

Figure 4.11 Topography causes a change in Moho depth (Archimedes' principle). The change in Moho depth itself has a gravity effect, at long wavelengths equal and opposite to the Bouguer effect.

variation in the depth to the base of the crust has its own semi-regional gravity effect and should be taken into account in any modelling that is undertaken (Chapin, 1998; Watts, 2001; Vixo and Connard, 2020).

Simplified isostatic corrections can be derived from topography alone. To first order, at long wavelengths, the isostatic effect is equal and opposite to the Bouguer effect. This is a simple application of Archimedes' principle (see Figure 4.11). In practice, it is often sufficient to take the Bouguer correction, upward continue it by (say) 32 km, reverse it and apply it as the Isostatic correction. This has the effect of cancelling the Bouguer correction at long wavelengths and displaying the geology to best advantage (see the case study in Section 4.11.1). It also helps explain the observation that the onshore Bouguer anomaly is typically negative. It is because we have not yet performed the necessary isostatic correction.

More rigorous calculation of isostatic corrections would also take account of the effect of large sedimentary basins and crustal rigidity. But this requires extra information that may not be available in greenfield areas. A more sophisticated approach which includes the Moho in the model and enforces isostatic balance is described by Vixo and Connard (2020). It MUST be remembered that isostatic equilibrium does NOT apply near plate active boundaries, be they subduction zones or rifts. Nor does it apply fully in regions recently relieved of glacial loading. They may be experiencing 'isostatic rebound'.

Data Acquisition Modes

Gravity acquisition has often been done in conjunction with seismic surveys, both onshore (shot-point gravity) and offshore (profile gravity), because there was a great saving in logistical (particularly survey) costs.

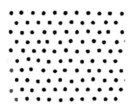

Figure 4.12 The optimal point gravity station layout to obtain best resolution with any given number of gravity stations per unit area is an equilateral triangular mesh.

It is becoming more common to conduct independent area-covering surveys, either on lines or (more recently) logically laid out grids, on the ground (point-based) or else airborne (line-based).

Survey Design

Survey design is directly based on the survey exploration objectives, but with the advent of satellite navigation and height measurement, there is no longer a strong advantage to collecting ground point gravity on widely spaced lines (aka 'shot-point gravity'). Modern surveys tend to use rationally laid out patterns of grid points. The station (and line) spacing determines the spatial resolution. The most space-efficient point layout pattern is an equilateral triangular mesh, as shown in Figure 4.12. The station spacing determines the spatial resolution of the survey.

Typical Applications of Gravity Surveys

Typical exploration uses of gravity survey data:

- Basin delineation and basin modelling
- Structure delineation (faults)
- Salt problems (location and delineation)
- Part of a pre-stack depth migration workflow to help constrain velocities
- Joint interpretation with seismic data. This helps the interpreter choose among different seismic interpretation possibilities.

Innovative Applications of Gravity Surveys

Absolute gravity surveys are now available at very high sensitivity. This eliminates the drift associated with conventional gravimeters and makes 4D gravity surveys available. This offers an obvious application to reservoir monitoring, as demonstrated by the monitoring of a waterflood enhanced gas production project on the Prudhoe Bay Gas Field in Alaska (Hare et al., 2008).

4.9.4 Magnetic Fields
The Deep Science

Magnetic fields follow the principles laid out by James Clerk Maxwell (encapsulated in Maxwell's equations). From a modern physics perspective, magnetic fields arise as the interactions between moving electric charges (or electric currents). Magnetic phenomena are in fact relativistic effects (a la Einstein), but it turns out that Maxwell's equations are fully compliant with relativity (despite their nineteenth-century formulation) and are therefore fully up to date. Beware practitioners who suggest they are more up to date than Maxwell.

Practical Consequences

The elementary magnetic source is a magnetic dipole (physically speaking it is a small current loop). As such it is a vector. And it gives rise to a set of magnetic 'lines of force' which form closed loops beginning and ending on the dipole. *That implies that any magnetic anomaly has as much 'low' as 'high' in it and any processing technique or interpretation approach needs to keep that firmly in mind. The 'low' is as important as the 'high'.*

Magnetic Field Units

The unit of magnetic field (strictly the magnetic flux density) most widely used in geophysics is the nanotesla or nT (formerly called gamma). The earth's magnetic field is between 25,000 nT (equatorial) and 50,000 nT (polar). The earth's magnetic field is weak. It is strong enough to orient a delicately balanced magnetised needle (a compass). We can measure that weak field to a few picotesla (say 1 ppm). The precision, stability and reliability of a modern geophysical magnetometer is a triumph of instrumentation, but that in turn requires painstakingly careful survey procedures to realise the full exploration benefit of the available sensitivity.

Magnetic Susceptibility Units

Magnetic susceptibility (χ) expresses the degree to which a material is capable of being reversibly magnetised by an external magnetic field. It is a dimensionless quantity, but there is a difference of definition between cgs-emu and SI, which gives rise to the relationship

$$\chi_{SI} = 4\pi \, \chi_{cgs \ emu.}$$

In this chapter we use SI units consistently.

The International Geomagnetic Reference Field and Time-Related Effects

The geomagnetic field arises largely from electric currents in the earth's core and is approximated by a geocentric dipole inclined at about 11° to the rotation axis. In consequence the geomagnetic field points up at the south dip pole, upward and northward in the Southern Hemisphere, horizontally and northward at the 'magnetic equator', downward and northward in the Northern Hemisphere, and directly downward at the north dip pole. The induced magnetic anomaly observed over a given geological structure therefore depends on location and can be quite different at different magnetic latitudes (Figure 4.13).

The 'main field' is modified by the effects of currents in the upper atmosphere, which give rise to a wide variety of time-related effects such as the 'diurnal',

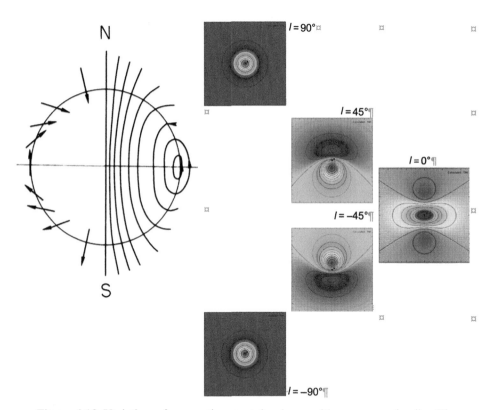

Figure 4.13 Variation of magnetic anomaly shape with geomagnetic dip. The variation in magnetic inclination (local dip) is shown on the left. The anomaly shape from a simple buried sphere exhibiting induced magnetisation is shown on the right. At either dip pole, the magnetic anomaly is a high with a diffuse surrounding low. At the magnetic equator, where the field is horizontal, it is a low over the source with flanking highs north and south. At intermediate inclinations we see a high–low pair, with the high towards the equator. Pole reduction simplifies this.

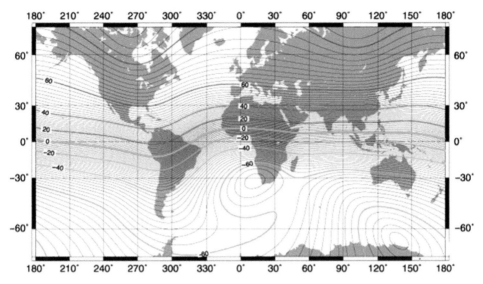

Figure 4.14 Variation of magnetic dip over the earth's surface (IGRF Inclination, year 2000). The green line is the 'magnetic equator', where the geomagnetic inclination (dip) is zero. The south dip pole is near the coast of Antarctica, south of Australia. The north dip pole is in the Canadian Arctic islands (Source: NOAA)

'magnetic storms', micropulsations and the equatorial electrojet. Dealing with these effects is the province of specialist contractors and consultants. The spatial (and temporal) variation of the strength and direction of the earth's main magnetic field is documented and released every 5 years as the International Geomagnetic Reference Field (IGRF; Figure 4.14). It is normal practice to subtract the IGRF value (at the survey date) from any survey data to remove gross regional effects to first order. The result should be termed the 'magnetic anomaly', the 'total field magnetic anomaly' or the 'IGRF residual' but is often misleadingly termed the 'total magnetic intensity' (TMI), which should, of course, be reserved for the actual measured total field value **before** IGRF subtraction.

The IGRF is also normally used to design pole reduction filters for any given location. It is available from NOAA from www.ngdc.noaa.gov/geomag/magfield-wist/.

Standard Corrections

Standard corrections applied to aeromagnetic survey data are

- Aircraft manoeuvre (figure of merit measure) and heading effects
- Diurnal and equatorial electrojet variations (base subtraction and/or levelling)
- IGRF (usually serves well as a regional-removal)
- Tie-line levelling

Data Acquisition Modes

Magnetic survey data can be collected on foot or from a vehicle (with special precautions). It is sometime collected offshore on seismic survey vessels using an extra streamer, but this is unpopular with crews and mag streamers are often 'lost'. Tailbuoy magnetometers are sometimes a good compromise. The best and cheapest data are collected by air, most often from a fixed-wing light aircraft with the sensor in a tail stinger. Variations use helicopters or towed 'birds'.

Survey Design and Specifications

Magnetic survey design is discussed by Reid (1980). Wide line spacing gives rise to undersampling and aliasing. Good survey data should be collected on lines spaced no wider than twice the height above significant magnetic sources (for <5% alias power). For detailed enhancements such as gradients, line spacing equal to height above sources is required in principle. It is also beneficial to fly magnetic surveys N–S rather than E–W, because the dipolar nature of magnetisation gives rise to anomalies that are more detailed N–S than E–W. A full discussion of survey specifications is given by Coyle et al. (2014). This GSC Open File is freely downloadable.

Typical Magnetic Anomalies (and Some Comments)

- Magnetic anomalies ALWAYS have both lows and highs associated with them. The anomaly from any isolated body will always sum to zero (it is one of Maxwell's equations). If you cannot see both the high and the low, you do not understand the anomaly.
- The shape of a magnetic anomaly depends not only on the geological source shape, but also on the local magnetic field direction (declination and inclination, that is, direction and dip). The direction is mostly northwards although a compass correction is necessary. The dip can be anything from straight up (at the south magnetic pole), through horizontal (at the magnetic equator) to straight down (at the north magnetic pole). In consequence the same geology can have quite different-looking magnetic anomalies, depending on the local magnetic field dip.
- Shallow structures give rise to anomalies with a small-scale 'rough' texture. Deep structures give rise to large-scale 'smooth' anomalies.
- Basic intrusions (and banded iron formation [BIF]) can give rise to anomaly amplitudes >1,000 nT. Sediments such as shales containing haematite (and perhaps a little biogenic or other magnetite) can show anomalies of 1 nT over faults.
- Beware of the effects of remanent magnetisation. We normally assume that magnetisation is 'induced', that is, parallel to the ambient geomagnetic field. This is most often true, but many rock masses retain a magnetisation associated

with their geological past, and the magnetisation may be in other directions. Where this is dominant, odd magnetisation signatures occur. Anyone who ignores this possibility (or fails to recognise it) is in danger of silly interpretations akin to seismic interpreters who fail to recognise 'ghosting'.

• Given the aforementioned oddities, magnetic interpretation is a specialist job.

4.10 Interpretive Techniques

Interpretive techniques have undergone huge development, largely in consequence of increased computer power. They fall into three groups:

• Interpretive enhancements
• Deconvolutions and depth estimators
• Modelling and inversion

4.10.1 Interpretive Enhancements

Interpretive enhancements are mostly implemented in the wavenumber domain employing fast Fourier transforms (FFTs). They include

• Horizontal gradients (which display ridges over contacts)
• Vertical gradients (which enhance the effects of shallow geology at the expense of deeper geology)
• Upward and downward continuation
• Pole reduction, equator reduction, pseudo-gravity (from magnetic data) and pseudo-magnetic displays (from gravity data) to centre geophysical effects over the causative geology
• Tilt and tilt derivatives (to display edges). 'Tilt' is the dip of the local field anomaly gradient. 'Tilt HD' is the horizontal gradient of Tilt. This filter shows ridges over structural edges (Verduzco et al., 2004). An example offshore Uruguay is shown in Figure 4.15.
• Pseudo-depth slicing exploits the fact that the log power spectral slope of a magnetic profile or grid is very simply related to the depth to source (Spector and Grant, 1970). It is not straightforward to extract a clean power spectrum free of edge effects, but if done correctly, a clean power spectrum will break up unto straight line segments, each characteristic of a source depth ensemble. This may be exploited by using spectral shaping filters to favour information from any particular source group at a consistent depth (Cowan and Cowan, 1993). A full separation is never possible (hence the 'pseudo-'), but significant enhancement of each depth group is certainly possible. The East Irish Sea Case Study in the text that follows uses this approach.

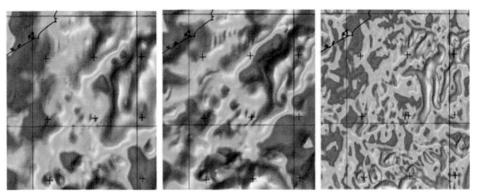

Figure 4.15 Enhancements of EMAG2 magnetic anomaly data offshore Uruguay (see earlier).
(Left) Pole reduction, centres anomalies over sources. (Centre) Tilt, produces a sharper image in place, with AGC (Right) Tilt HD, shows ridges above structural edges.

4.10.2 Deconvolutions and Depth Estimators

Deconvolutions and depth estimators typically exploit the field curvature to estimate depth. This is a fertile and fast-moving field. Operations include

- Spectral slope (Spector and Grant, 1970) may be used on windowed portions of a grid to estimate depth to basement (or other significant magnetic horizons) in a piecewise manner. Extracting a clean windowed log power spectrum requires specialist software, but this is my personal favourite basement depth estimator. It is used in the Lake Edward (Uganda) Case Study in the text that follows.
- Werner and Euler deconvolution (Werner, 1955; Hartman et al., 1971; Reid et al., 1990; Mushayandebvu et al., 2001; Reid et al., 2014)
- Naudy depth (Naudy, 1971)
- Tilt depth (depth derived from rate of change of tilt; Salem et al., 2008, 2010)
- Source Parameter Imaging™ (iSPI; Smith et al., 1998)
- GridSLUTH (Source Location Using Total Field Homogeneity; - Smith et al., 2012). An example is shown in Figures 4.16 and 4.17.

Of the aforementioned processes, Werner deconvolution and Naudy depth imaging were very widely applied to profile data and are still applied if the only data available are widely spaced profiles, but have been somewhat superseded by the others, which operate on grids. Euler is probably the most widely used because a commercial software tool is readily available. Tilt is a very good technique for extracting structural edges, but Tilt Depth is perhaps available only from some specialist contractors. GridSLUTH and iSPI make the least assumptions and are, on those grounds, preferable, but they are currently more widely used in the minerals exploration sector. Petroleum work would benefit from their wider use.

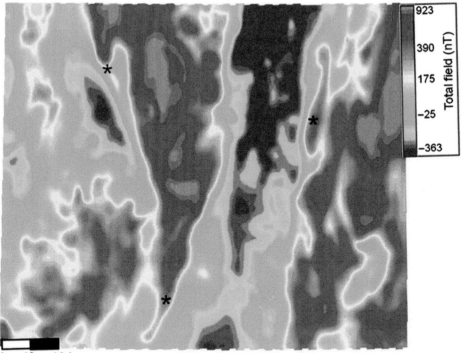

Figure 4.16 Aeromagnetic data, Saskatoon area. (Smith et al., 2012)

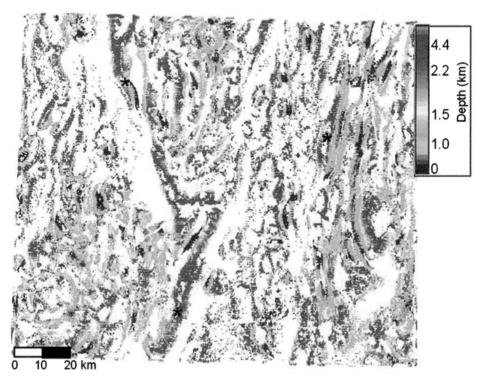

Figure 4.17 Depth to basement using GridSLUTH method, Saskatoon area. (Smith et al., 2012)

My personal favourite magnetic basement depth estimator is the power spectrum slope, but this requires specialist software.

4.10.3 Modelling and Inversion

Modelling and inversion are major subjects beyond the scope of this chapter, but it is now possible to create sophisticated geologically constrained models of most scenarios of interest to investigate and match the geology to the gravity, magnetic or other geophysical signatures. Some examples are shown in the Case Studies section in the text that follows.

Most of these techniques require proprietary software or services available from specialist contractors. Most of them have been described in the literature and are open to development by anyone with sufficient expertise and diligence. There are good overview papers on both magnetic (Nabighian et al., 2005a) and gravity survey practice and interpretation (Nabighian et al., 2005b).

Typical Applications of Magnetic Surveys

Typical exploration uses:

- Depth to basement (Vacquier et al., 1951)
- Basement and basin structure
- Major fault delineation
- Detection and delineation of intrusives and extrusives
- Fault connection within sediments, in HRAM surveys (Saad 2018a, b)

4.11 Case Studies

The case studies illustrate the diverse benefits that can flow from use of gravity and magnetic studies integrated into the exploration programme. In order of appearance they show

1. Basin delineation offshore, using public domain data available for anywhere offshore (Guinea, West Africa)
2. Simple basin and structural delineation using purposely collected airborne gravity and magnetic, seismic, seep and drill-hole data (Lake Edward, Uganda)
3. Magnetic delineation of shallow sedimentary structures (North Sea)
4. Detailed magnetic and seismic mapping of faults directly controlling hydrocarbons (and a hint of direct hydrocarbon indication [DHI] – East Irish Sea)
5. Seismic-controlled inversion of gravity to delineate sedimentary structure (North Sea) and
6. Detailed inversion of gravity data for basin structure, with a full suite of geological constraints (onshore Western Australia)

4.11.1 Offshore Satellite Gravity and Isostatic Residuals, Republic of Guinea

The images shown below are derived from data that are available with free usage permission on the internet. The exploration purpose of the study was to delineate possible offshore basins, as a first step in determining the prospectivity of the continental shelf offshore Republic of Guinea.

The freely available data sets concerned are offshore free air gravity, bathymetry and adjacent onshore topography.

The bathymetry and topography data were obtained from www.gebco.net (GEBCO 2020). Adjacent onshore topography was obtained from https://earthexplorer.usgs .gov. Merged bathymetry and topography are shown in Figure 4.18.

Free air gravity derived from satellite altimetry was obtained from http://topex .ucsd.edu/cgi-bin/get_data.cgi.

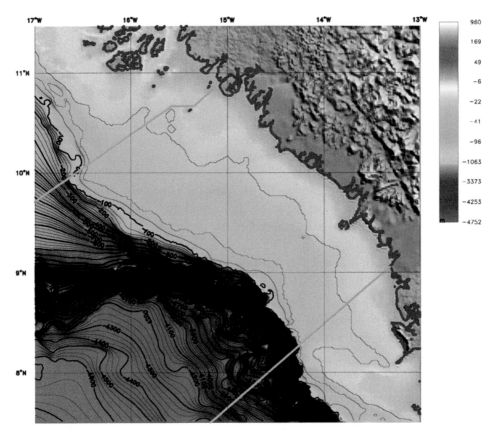

Figure 4.18 Bathymetry offshore, topography onshore Republic of Guinea, West Africa. Orange outline marks Guinean territorial waters. Note extreme shelf-edge drop-off.

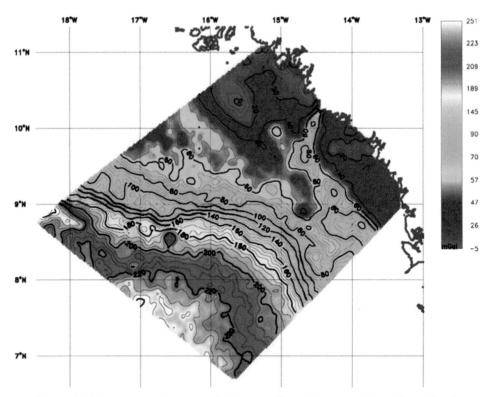

Figure 4.19 Bouguer gravity anomaly from satellite altimetry, offshore Republic of Guinea. Note the strong gradient associated with the shelf edge. This gradient is a consequence of isostasy.

The available satellite gravity is the free air anomaly. This is directly affected by variations in water depth on continental margins, and this effect may be treated using the Bouguer correction, to arrive at the Bouguer anomaly (Figure 4.19).

It is also affected indirectly by isostasy at continental margins. The isostasy gives rise to crustal thinning, which has a gravitational effect. We can estimate the isostatic effect by applying Archimedes' principle and assuming Airy isostasy and isostatic equilibrium. This is reasonable at a passive margin. The result of applying Bouguer and isostatic corrections is the 'isostatic residual anomaly' (Figure 4.20).

The isostatic residual should respond directly to density changes within the crust. Offshore Republic of Guinea, there is an appreciable continental shelf and then a very steep bathymetry drop-off to the deep ocean. On the continental shelf, several separate sedimentary basins are clearly visible, one open to the north in Guinea–Bissau waters. There is also a strong hint of shelf edge sedimentation.

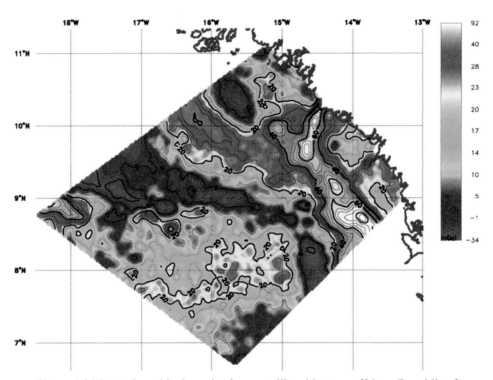

Figure 4.20 Isostatic residual gravity from satellite altimetry, offshore Republic of Guinea. Note the basins in the near offshore and the possible deep sediments at the continental edges. The isostatic residual has removed gross gradients. The basin architecture is much clearer.

This approach may be applied anywhere offshore (away from tectonic plate edges), and is an indispensable start point for anyone contemplating further petroleum exploration in any shelf area.

4.11.2 Fault-Bounded Basin and Internal Structure Delineation: Lake Edward

The East African Rift lakes are known to host significant petroleum occurrences. Drilling in the Lake Albert Basin even had to cope with a 'gusher' (E. Kashambuzi, pers. comm.). Lake Edward showed similar promise, encouraged by known static oil seeps observed on the lake surface. As part of an integrated exploration programme that also involved geological mapping, onshore seismic work and drilling, Dominion Petroleum commissioned an airborne gravity and magnetic survey over Lake Edward. It was operationally challenging, given the surrounding mountains. The free air gravity anomaly over the lake exceeded 140 mGal. Even the Bouguer gravity anomaly exceeded

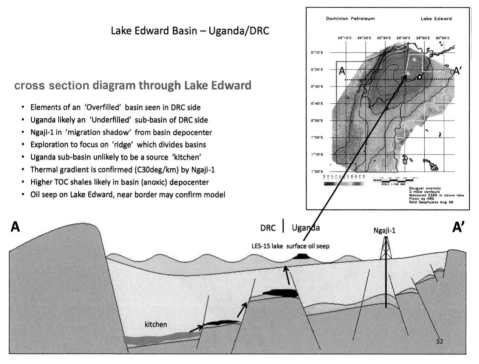

Figure 4.21 Bouguer gravity anomaly over and cross section model through Lake Edward.

80 mGal, in accordance with expectations of a thick accumulation of young sediments. The gravity data were used for both fault delineation and for profile modelling constrained by

1. The Ngaji-1 well
2. Seismic data on the eastern margin of the rift valley
3. Basement depth estimation (by the spectral slope method; see Section 4.10.2) within the graben using the magnetic data observed simultaneously.

The result is shown as Figure 4.21. The LES-15 oil seep is also an observation. The depicted reservoirs are speculative.

In addition to the cross section model, gravity gradient analysis was used to delineate faulting within the basin sediments. The interpreted faults were well supported by the sparse seismic data and provided vital structural detail over the entire basin. Part of this is shown in Figure 4.22.

The licences have now lapsed, and the rights are held by the Government of Uganda. We acknowledge their permission to show these data.

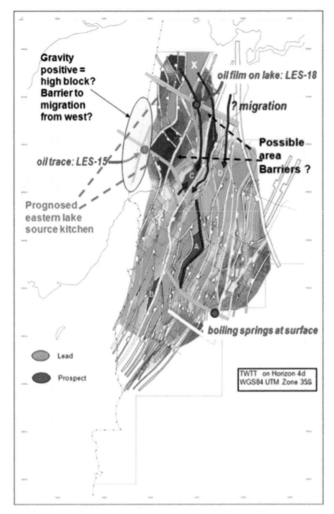

Figure 4.22 Lake Edward structural interpretation from gravity gradient enhancement, guided by limited onshore seismic data on the east shore.

4.11.3 Sedimentary Structures, North Sea Quaternary Tunnel Valleys

The increase in resolution and sensitivity of aeromagnetic surveys makes it possible, in favourable situations, to delineate shallow structure within petroleum-bearing sediments. This case study has been included here because it is one of the few fully documented examples in the refereed literature (Brahimi et al., 2020).

High-resolution aeromagnetic data were used to map late Quaternary subglacially cut tunnel valleys in the Central Viking Graben (Figure 4.23).

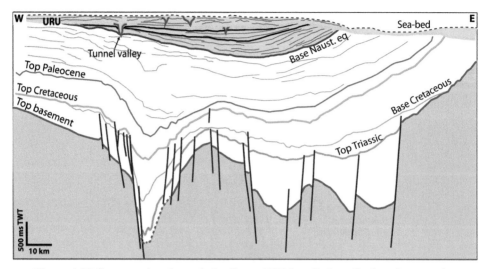

Figure 4.23 Cross section through the Central Viking Graben (for location, see the original article). Shallow Quaternary tunnel valleys are marked.

Low-noise, high-resolution aeromagnetic data were available. Low-amplitude, short-wavelength effects are not apparent in the primary anomaly data, even though the area is also crossed by several pipelines and cables (Figure 4.24a). Short-wavelength features may be enhanced by appropriate high-pass filtering. A vertical derivative of order 1.7 was chosen and is shown as Figure 4.24b.

We see clear imaging of the tunnel valleys, as well as the pipelines and submarine cables.

This example illustrates that, with sufficiently high-resolution, low-noise aeromagnetic data, subtle shallow sedimentary features may be mapped using aeromagnetic data. Subtle features such as these can probably only be seen in the top 1,000 m of the section, but it depends on local mineralogy. The subject is further discussed by Saad (2018a, 2018b).

4.11.4 Sediment Fault Mapping: East Irish Sea

In favourable circumstances, high-resolution, high-sensitivity aeromagnetic data can yield an extraordinary amount of information, directly relevant to petroleum exploration and production. This is one such example, where aeromagnetic data combined with legacy seismic data made it possible to generate a dense structural framework over a large area, including known and prospective fields. There was also a hint of DHI.

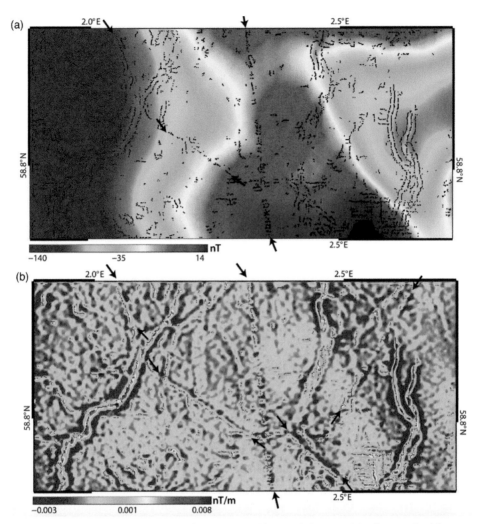

Figure 4.24 (a) Magnetic anomaly over a small part of the area. Pipelines and cables are marked with arrows. Black linework follows ridges in the vertical derivative depicted in (b). (b) Vertical derivative (order 1.7) of the magnetic anomaly. Several tunnel valleys are clearly shown, as are the effects of pipelines and cables.

A joint venture between Robertson Research International and World Geoscience Corp. (both now parts of CGG) flew a non-exclusive low-level, high-sensitivity HRAM survey over the East Irish Sea in 1992, in an attempt to resolve sedimentary structure. The survey coverage and known structure are shown in Figure 4.25. The survey overflew the existing Morecambe Bay gas field, which had an original gas content of 700 tcf, and the then developing Hamilton and Douglas Fields.

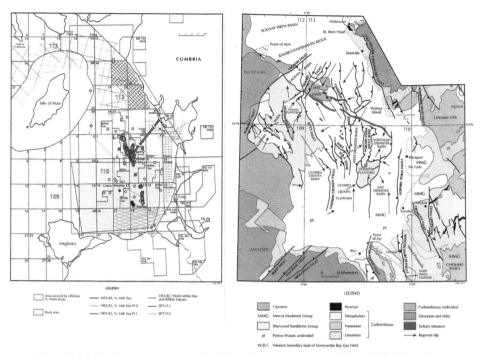

Figure 4.25 Project coverage and existing known structure, East Irish Sea. (From Reid, 1994)

The existing seismic data over the area consisted of a 10-km grid of 2D lines. They showed a significant amount of shallow fault structure, but with the wide spacing, any fault connection was necessarily speculative.

The reservoir rock in the area is the Triassic Sherwood Sandstone, overlain and sealed by the Mercia Mudstone. The aeromagnetic survey might reasonably be expected to see subtle magnetic signatures from faults breaking this contact, because the Mercia Mudstone should be weakly magnetic. A typical depth to the gas–water contact is about 1,000 m.

The survey was flown with flight-lines oriented 020°– 200°at a spacing of 400 m and perpendicular tie-line spacing of 1,200 m and a sea surface clearance of 80 m. Every effort was made to get the survey noise as low as possible (below 0.1 nT). The data were subjected to power spectrum analysis and pseudo-depth slicing (see Section 4.10.1). The actual slice depths employed are proprietary information, but a combination of slices 1, 2 and 3 is shown in Figure 4.26, along with a simplified fault interpretation. The data are available from CGG Geostore (https://bit.ly/2TCC2gX)

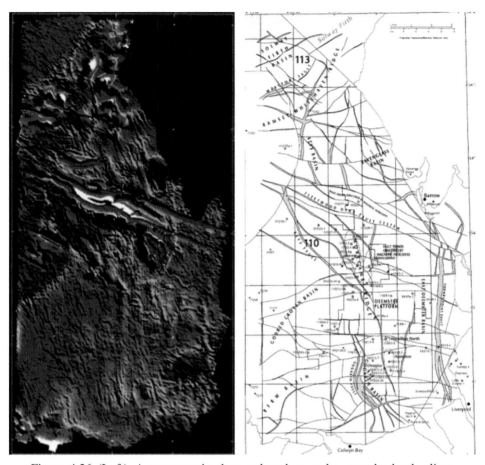

Figure 4.26 (Left) Aeromagnetic data reduced to pole; pseudo-depth slices 1–3. The strongest magnetic signature is from the Fleetwood Dyke and the sill fed from it. The fine lower-amplitude linear structure is probably faulting at the Sherwood Sandstone–Mercia Mudstone contact. (Right) Structural elements interpreted from aeromagnetic data and existing regional seismic coverage.

The interpretive work identified faults in the seismic sections and used the magnetic linears to carry out the fault connection.

In addition to meeting its objectives, the study achieved two notable spin-off results.

1. The interpretation identified cross-cutting faults within the Morecambe Bay Gas Field that are not imaged by the existing 2D seismic data because they run parallel to the seismic lines. The cross-cutting faults are of direct relevance to

production. Such faults would have been imaged by modern 3D seismic surveying, but at very much greater cost.

2. The shallowest magnetic data depth slice showed a subtle (probably remanent) magnetic low corresponding roughly to the outline of the gas field. Further work suggested that this is in fact the gas–water contact (GWC), or an earlier paleo-GWC (Morgan, 1998). This is plausible, because bacteria such as *Geobacter metallireducens* are known to metabolise organic compounds (such as petroleum) while reducing haematite to magnetite (Lovley et al., 1987), so we are not surprised if a gas–water contact had a magnetic signature. This is our 'DHI'.

4.11.5 Seismic-Controlled Gravity Inversion: Central North Sea

This case study briefly describes a comprehensive regional synthesis which integrates modern long-offset seismic data with dense grids of gravity and magnetic data to achieve a regional-scale synthesis of Devonian, Carboniferous and Early Permian basin development beneath the UK Central North Sea. It also seeks to identify possible accumulations of pre-Permian Carboniferous rocks which may contain coal-measure source rocks, opening up the possibility of a carboniferous petroleum system. Those with a close interest in the topic are referred to the source paper (Milton-Worssell et al., 2010).

The study created a model which included everything from top-Moho to the sea surface. Constraints included the following elements:

1. The top-Moho was estimated using a previous refraction seismic study.
2. Permian and younger strata were constrained by information from a wide variety of sources but included comprehensive seismic and well data from an area which has been well explored down to Base Permian. These shallower layers were represented by a 'density cube' derived from a depth-converted version of the 'velocity cube' over the entire area.
3. Some parts of the top basement were delineated by long-offset seismic data.

That left the remainder of the top-basement free to estimate, using gravity and magnetic data. It proved necessary to account for inhomogeneities within basement, by explicitly incorporating three lower density granites. The model for top-basement to Moho is shown as Figure 4.27.

The main benefit of this study was to show that there is plenty of room between base Permian and basement for a substantial Carboniferous section, with source rocks likely to be at a suitable maturity, strongly supporting the

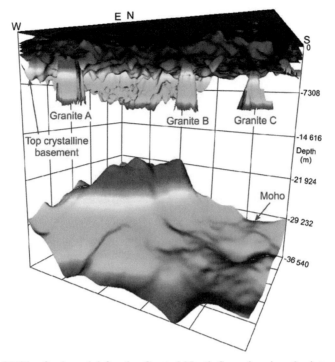

Figure 4.27 The final model for the Central North Sea, showing the interval from Moho to top-basement. The Moho was derived from seismic refraction data. The top basement was generated by inversion of the gravity data, with the insertion of three granitic bodies into the basement. The final model is shown from a higher perspective in Figure 4.28, along with a seismic section illustrating the shallower strata.

suggestion of a substantial Carbonifous petroleum system in the Central Graben.

4.11.6 3D Geologically Constrained Modelling, Merlinleigh Sub-basin, Western Australia: Purpose and Structure of the Study

The exploration value of this work is in the clear definition of the basin architecture and of the structural controls on that architecture. The work was undertaken on behalf of the State of Western Australia and was intended to provide a more certain exploration framework for intending petroleum explorers (Burney et al., 2014).

This case study shows the integration of geological, seismic, gravity and magnetic data to create a model which honours all the geological observations and

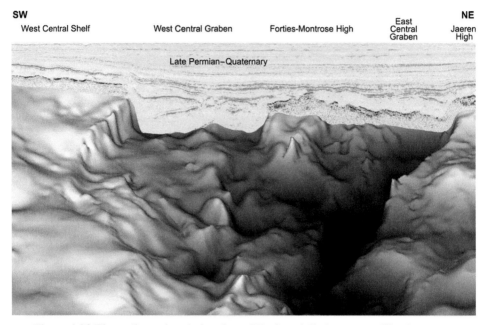

Figure 4.28 Three-dimensional view from SE of modelled top crystalline basement surface intersecting a 2D long-offset seismic profile across the southern part of the study area, which traverses the West-Central Shelf, the West-Central Graben, the Forties-Montrose High, the East-Central Graben, and ends on the Jaeren High with its Devono-Carboniferous core. The sub-Zechstein 'composite layer' is uncoloured on the seismic profile.

uses the potential fields data to constrain the basement depth. In that sense, it is similar to the North Sea example shown in Section 4.11.5.

It expands on the previous example in using the geological and seismic data sets as formal constraints on the inversion, which also takes the gravity and magnetic data into account. This is possible via the 3D geological and geophysical modelling package *GeoModeller*©, a system originally developed by the Bureau de Recherches Géologiques et Minières (BRGM – French Geological Survey), and now available commercially from Intrepid Geophysics.

Geological Setting and Geophysical Data Processing

The Merlinleigh sub-basin is located in the onshore Carnarvon Basin, Western Australia. It is an epicratonic rift basin, filled with shallow marine sediments covering the age range Late Silurian to Mid Carboniferous. The petroleum potential is not yet established.

As part of the Petroleum Initiatives program conducted by the Geological Survey of Western Australia (GSWA) high-resolution aeromagnetic and

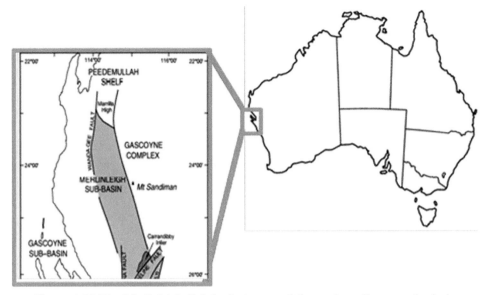

Figure 4.29 The Merlinleigh Sub-basin is part of the onshore Carnavon basin in Western Australia. The region's Proterozoic basement is overlain by sediment up to 8,000 m thick.

semi-detailed helicopter-supported gravity surveys were conducted in 1995 to assist with the structural interpretation of the Merlinleigh Sub-basin.

3D Geological Modelling

A 3D geological model of the Merlinleigh sub-basin was created using GeoModeller[©]. The software uses a 3D probability model calculated from imported geological data. Data can include drill holes, mapped surface lithology, contact and orientation data and cross sections.

For this project the terrain was modelled using the SRTM 90m data from CGIAR-CSI (Consortium for Spatial Information: https://cgiarcsi .community).

The geology of the Merlinleigh basin was created and constrained using geological surface maps and interpreted seismic cross sections from Lasky et al. (1998). Some of the geological constraints and their digital implementations are depicted in Figure 4.33. The modelling system imposes geologically reasonable constraints such as an enforced stratigraphic sequence.

The constraining cross sections were derived from the existing sparse seismic coverage (Figure 4.34).

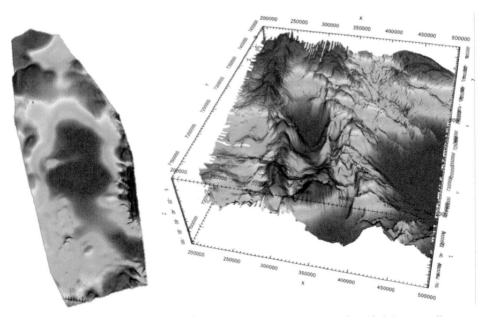

Figure 4.30 Aeromagnetic (left) and ground Bouguer gravity (right) anomalies over the Merlinleigh Sub-basin of the Carnavon Basin, Western Australia.
Multiscale edge extraction was applied separately to both magnetic and gravity data. This employs upward continuation, horizontal gradient and ridge following to generate a representation of the significant edges in the geological structure and their approximate depths. The edges from magnetic data are shown here. The edges from both analyses were taken into the later geological modelling stage.

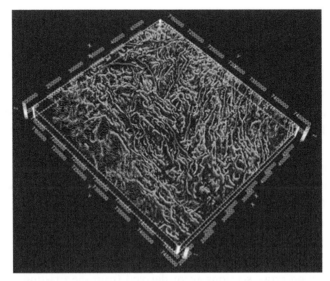

Figure 4.31 Multiscale edge extraction and shallow-deep separation of gravity effects, Merlinleigh Sub-basin.

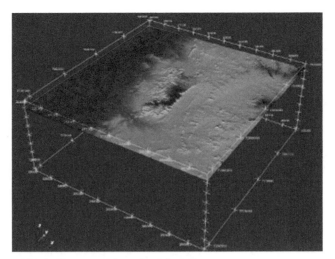

Figure 4.32 SRTM Iimported into GeoModeller.

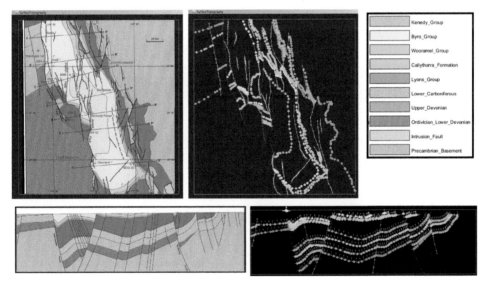

Figure 4.33 Imported geological information digitised from maps and cross sections and imposed stratigraphic pile.

The final constrained inversion fitted the gravity data, using the whole model. The final result is shown in Figure 4.35. It was found necessary to include a dyke-like basement feature, aligned with the Wandagee Fault.

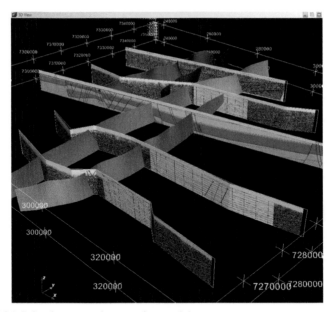

Figure 4.34 Seismic constraints on the model.

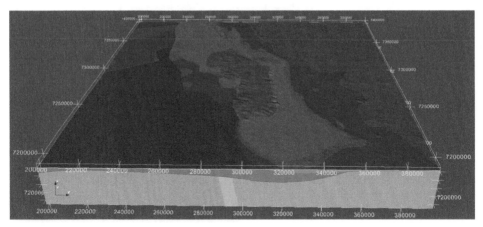

Figure 4.35 Final inverted model, fitting all geological and geophysical constraints. Note one dyke-like intra-basement feature, aligned with the Wandagee Fault, which was found to be necessary.

Since the completion of this study, the Geological Survey of Western Australia has used these results directly in its resource and maturity modelling, as part of the data package to encourage further commercial exploration of this very promising but under-explored structure.

4.12 Application Environments/Discussion

4.12.1 Situations/Applications Where the Technology Works Well

Gravity and magnetic measurements are most useful in situations where there is lateral change (contrast) in density or magnetisation. They are good at 'seeing' normal faults and near-vertical contacts and not so good at strike-slip faults unless they have very considerable movement on them.

Horizontal contacts are best seen with borehole and reflection seismic measurements.

The 'depth information' in magnetic fields essentially lies in the curvature of the field.

The main difference between gravity and gravity gradient survey measurements and their applications is that gravity gradients detect and preserve information about detail and short wavelengths. Their reliability decreases with wavelength and becomes zero at scales longer than the flight-line length, Gravity data detect and preserve longer wavelengths. If you want the basin configuration and you care about the relative depth of adjacent sub-basins, you need gravity. If you want the best possible delineation of shallow normal faults, or detailed shallow salt structures, and care about little else, you need gravity tensor gradient, and a closer line spacing (with its increased cost per square kilometre). Some survey companies offer them together, which gives the best of both.

4.13 Future Developments

4.13.1 Technology

An excellent overview of petroleum applications of gravity and magnetic fields is to be found in the SEG/AAPG volume *Geologic Applications of Gravity and Magnetics: Case Histories* (Gibson and Millegan, 1998)

We look forward to

- Improved gravity gradiometers
- Fly-able absolute gravimeters and gradiometers based on either free-fall measurements or 'cold atom' gravimeters
- Availability of cheaper portable gravimeters based on microfabricated accelerometers
- Increased use of UAV platforms for magnetic (and perhaps gravity) surveys (Figure 4.36)
- Incremental improvements in magnetic sensitivity and reductions in line spacing and flying height to improve resolution and detection of sedimentary features

Figure 4.36 UAV aeromagnetic survey system (Stratus Aeronautics).

- Wider availability of magnetic tensor gradient systems
- Further improvements in satellite measurements
- 4D gravity, already here but not yet widely used

4.13.2 Applications

We expect to see

- Increased use of magnetic surveys for delineating structure in sediments, because it is very cost-effective, where applicable
- Advanced interpretation algorithms taking advantage of tensor data to estimate dips and improve resolution

- Increased use of absolute gravity measurements (using falling weight systems) for 4D surveys doing water flooding enhanced gas production (Hare et al., 2008) and reservoir monitoring
- Increased borehole gravity and borehole gravity gradient, including horizontal wells
- Greater integration of results from seismic, EM and potential fields methods (e.g. Moorkamp et al., 2013)

4.14 Supplementary Material

4.14.1 Geoidal versus Ellipsoidal Height

By convention and established geodetic practice, all gravity work is done with elevations referenced to the geoid, which is the height that would be derived by conventional survey levelling, starting at mean sea level. So, all aforementioned heights are geoidal heights. Satellite navigation systems (e.g. GPS) actually generate heights above the WGS-84 spheroid ('spheroidal heights'). These may be up to 100 m different (Figure 4.37). Most GPS systems can output geoidal heights instead, using built-in 'geoid reference models'. These are good enough

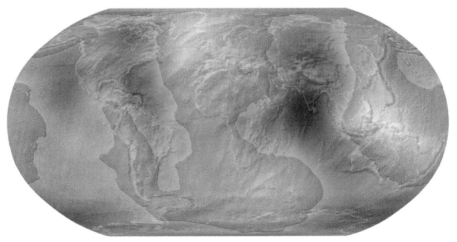

Red areas are above the idealised ellipsoid; blue areas are below.

−107.0 m   0 m   +85.4 m

Figure 4.37 Deviation of the Geoid from the idealised figure of the earth (the difference between the EGM96 geoid and the WGS84 reference ellipsoid). (Source: NASA)

for many purposes. For strict correctness, a gravity 'anomaly' of any type is referred to the geoid, whereas a gravity 'disturbance' of the same type is referred to the ellipsoid. The difference between the two affects only the longest wavelengths, so it seldom affects local detail, but it is wise to be clear which you are using and not mix them.

4.14.2 What Do We Mean by 'Vertical'?

In a similar way, the 'vertical' is defined by the direction of the local gravity field, as it might affect a plumb-bob, so it is everywhere perpendicular to the geoid. There is therefore no possibility of a horizontal gravity anomaly. It is zero by definition. But there is a possibility of a horizontal gravity *gradient* anomaly.

Nevertheless, by reference to the ellipsoid or astronomically based direction systems, there is significant 'deflection of the vertical', typically on the edges of major gravity anomalies. This has been measured (previously by reference to star sightings) by geodetic surveyors since the days of Bouguer himself.

It is therefore also possible (and useful) to measure horizontal gravity anomalies (or, more correctly, 'disturbances') from airborne gravity systems that typically use gyro-stabilised platforms as their local orientation reference. Such measurements have been demonstrated using currently available airborne survey systems but are not yet used routinely. They would be largest and most useful in mineral-prospecting situations.

4.15 Unit Conversions and Typical Parameter Values

Table 4.5 *Unit conversions in potential fields*

Parameter	Users	Unit	Size in SI base units
Density	Industry	$g \cdot cm^{-3}$	$10^3 \text{ kg} \cdot m^{-3}$
Gravity anomaly	Industry	mGal (milligal)	$10^{-5} \text{ m} \cdot s^{-2}$
Gravity anomaly	Industry	µGal (microgal)	$10^{-8} \text{ m} \cdot s^{-2}$
Gravity anomaly	Academe	gu (gravity unit)	$10^{-6} \text{ m} \cdot s^{-2}$
Gravity gradient	Industry	E (Eötvös)	10^{-9} s^{-2}
Magnetic anomaly	Industry	nT	$10^{-9} \text{ T (tesla)}$
Magnetic anomaly	Industry	pT	10^{-12} T
Angle	Industry	° (degree)	$0.017453 \ (\pi/180) \text{ radians}$
P-wave velocity	Industry	ft/s	$0.3048 \text{ m} \cdot s^{-1}$

Table 4.6 *Typical parameter values*

Parameter	Typical	Comment
Density	$1.0–3.0 \text{ g} \cdot \text{cm}^{-3}$	Fresh water – mafic basement
Gravity anomaly	Up to 150 mGal	Recent grabens (East African Rift)
Gravity noise	10 μgal	High precision local ground survey
Gravity noise	0.1 mGal	Good petroleum ground survey
Gravity noise	1 mGal	Typical airborne gravity
Absolute gravity	981,000 mGal	Absolute gravity value
Gravity gradient	3,086 E	Free air vertical gradient
Gravity gradient noise	2–10 E	Typical airborne FTG survey
Magnetic field	25,000–50,000 nT	Total magnetic field
Magnetic noise	0.1 nT (100 pT)	Good airborne magnetic survey
Magnetic mis-tie	<1 nT	Good airborne magnetic survey

References

Bonvalot, S., Balmino, G., Briais, A., et al., 2012. *World Gravity Map. Commission for the Geological Map of the World*. Paris: BGI-CGMW-CNES-IRD.

Brahimi, S., Le Maire, P., Ghienne, J. F. and Munschy, M., 2020. Deciphering channel networks from aeromagnetic potential field data: The case of the North Sea Quaternary tunnel valleys. *Geophysical Journal International*, **220**, 1447–62. DOI: 10.1093/gji/ggz494.

Braitenberg, C., 2014. Exploration for tectonic structures with GOCE in Africa and across-continents. *International Journal of Applied Earth Observation and Geoinformation*, **35**(Pt A), 88–95. DOI: 10.1016/j.jag.2014.01.013.

Burney, C., FitzGerald, D. and Zengerer, M., 2014. Non-seismic geophysical modelling methods for realistic characterisation of 3D geology in greenfields exploration. Poster paper presented at the 38th Indonesian Petroleum Association (IPA) Annual Convention and Exhibition, Jakarta, 21–23 May 2014.

Chandler, V. W. and Lively, R. S., 2015. 2011 Rock properties database: Density, magnetic susceptibility, and natural remanent magnetization of rocks in Minnesota. Retrieved from the Data Repository for the University of Minnesota, DOI: 10.13020/D63S3D.

Chapin, D., 1998. The isostatic gravity residual of onshore South America: Examples of the utility of isostatic gravity residuals as a regional exploration tool. In *Geologic Applications of Gravity and Magnetics: Case Histories*, 34–6. SEG Reference Series No 8. AAPG Studies in Geology No. 43. Tulsa, OK: Society of Exploration Geophysicists & American Association of Petroleum Geologists.

Cowan, D. R. and Cowan, S., 1993. Separation filtering applied to aeromagnetic data. *Exploration Geophysics*, **24**(4), 429–36. DOI: 10.1071/EG993429.

Coyle, M., Dumont, R., Keating, P., Kiss, F. and Miles, W., 2014. Geological Survey of Canada aeromagnetic surveys: Design, quality assurance and data dissemination. Geological Survey of Canada Open File 7660.

Gardner, G. H. F., Gardner, L. W. and Gregory, A. R., 1974. Formation velocity and density: The diagnostic basics for stratigraphic traps. *Geophysics*, **39**(6), 770–80. DOI: 10.1190/1.1440465.

GEBCO. 2020. General Bathymetric Chart of the Oceans. GEBCO Compilation Group (2020) GEBCO 2020 Grid. DOI: 10.5285/a29c5465-b138-234d-e053-6c86abc040b9.

Geoid. Wikipedia article. http://en.wikipedia.org/wiki/Geoid

Gibson, R. I. and Millegan, P. S. (eds.), 1998. *Geologic Applications of Gravity and Magnetics: Case Histories*. SEG Reference Series No. 8. AAPG Studies in Geology No 43. Tulsa, OK: Society of Exploration Geophysicists & American Association of Petroleum Geologists.

Hammer, S., 1983. Airborne gravity is here! *Geophysics*, **48**(2), 213–23. DOI: 10.1190/1.1441460.

Hare, J. L., Ferguson, J. F. and Brady, J. L., 2008. The 4D microgravity method for waterflood surveillance. Part IV: Modelling and interpretation of early epoch 4D gravity surveys at Prudhoe Bay, Alaska. *Geophysics*, **73**(6), WA173–WA180. DOI: 10.1190/1.2991120.

Hartman, R. R., Friedberg, J. L. and Teskey, D. J., 1971. A system for rapid digital aeromagnetic interpretation. *Geophysics*, **36**(5), 891–918. DOI: 10.1190/1.1440223.

Lasky, R. P., Mory, A. J., Ghori, K. A. R. and Shevchenko, S. I.., 1998. Structure and petroleum potential of the southern Merlinleigh sub-basin, Carnarvon Basin, Western Australia, Report 61, Geological Survey of Western Australia.

Lovley, D. R., Stolz, J. F., Nord, G. L. Jr. and Phillips, E. J. P., 1987. Anaerobic production of magnetite by a dissimilatory iron-reducing microorganism. *Nature*, **330**(6144), 252–4. DOI: 10.1038/330252a0.

Milton-Worssell, R., Smith, K., McGrandle, A., Watson, J. and Cameron, D., 2010. The search for a Carboniferous petroleum system beneath the Central North Sea. In B. A. Vining and S. C. Pickering (eds.), *Petroleum Geology: From Mature Basins to New Frontiers: Proceedings of the 7th Petroleum Geology Conference*, 57–75, London: Geological Society. DOI: 10.1144/0070057.

Minty, B. R. S., 2011. Airborne geophysical mapping of the Australian continent. *Geophysics*, **76**(5), A27–A30. DOI: 10.1190/geo2011–0056.1.

Moorkamp, M., Roberts, A. M., Jegen, M., Heincke, B. and Hobbs, R. W., 2013. Verification of velocity-resistivity relationships derived from structural joint inversion with borehole data. *Geophysical Research Letters*, **40**(14), 3596–601. DOI: 10.1002/grl.50696.

Morgan, R., 1998. Magnetic anomalies associated with the North and South Morecambe Field, U.K. In *Geologic Applications of Gravity and Magnetics: Case Histories*, 85–91. SEG Reference Series No 8. AAPG Studies in Geology No. 43. Tulsa, OK: Society of Exploration Geophysicists & American Association of Petroleum Geologists.

Mushayandebvu, M. F., van Driel, P., Reid, A. B. and Fairhead, J. D., 2001. Magnetic source parameters of two-dimensional structures using extended Euler deconvolution. *Geophysics*, **66**(3), 814–23. DOI: 10.1190/1.1444971.

Nabighian, M. N., Grauch, V. J. S., Hansen, R. O., et al., 2005a. The historical development of the magnetic method in exploration. *Geophysics*, **70**(6), 33ND–61ND. DOI: 10.1190/1.2133784.

Nabighian, M. N., Ander, M. E., Grauch, V. J. S., et al., 2005b. The historical development of the gravity method in exploration. *Geophysics*, **70**(6), 33ND–89ND. DOI: 10.1190/1.2133785.

Naudy, H., 1971. Automatic determination of depth on aeromagnetic profiles. *Geophysics*, **36**(4), 717–22. DOI: 10.1190/1.1440207.

Pawlowski, R., 2020. Years of Arabian Peninsula gravity exploration by Chevron and its legacy companies, including discovery of the Ghawar and Burgan super-giants. *The Leading Edge*, **39**(4), 279–83. DOI: 10.1190/tle39040279.1.

Reid, A. B., 1980. Aeromagnetic survey design. *Geophysics*, **45**(5), 973–6. DOI: 10.1190/1.1441102.

Reid, A. B., 1994. High resolution aeromagnetic data: New tricks for an old dog! Poster paper at GEO'94, Middle East Geoscience Conference, Bahrain, 1994.

Reid, A. B., Allsop, J. M., Granser, H., Millett, A. J. and Somerton, I. W., 1990. Magnetic Interpretation in three dimensions using Euler Deconvolution. *Geophysics*, **55**(1), 80–91. DOI: 10.1190/1.1442861.

Reid, A. B., Ebbing, J. and Webb, S. J., 2014. Avoidable Euler errors: The use and abuse of Euler deconvolution applied to potential fields. *Geophysical Prospecting*, **62**(5), 1162–8. DOI: 10.1111/1365-2478.12119.

Reynolds, R. L., Fishman, N. S., Wanty, R. B. and Goldhaber, M. B., 1990, Iron sulphide minerals at Cement oilfield, Oklahoma: Implications for magnetic detection of oilfields. *Geological Society of America Bulletin*, **102**, 368–80. DOI: https://bit.ly/2M9X4zo.

Saad, A., 2018a. Sedimentary magnetic anomalies. Part 1: The validity of short-wavelength, low amplitude SEDMAG anomalies. *The Leading Edge*, **37**(10), 774–9. DOI: 10.1190/tle37100774.1.

Saad, A., 2018b. Sedimentary magnetic anomalies. Part 2: Examples, sources and geologi9 origin of SEDMAG anomalies. *The Leading Edge*, **37**(10), 830–7. DOI: 10.1190/tle37110830.1.

Salem, A., Williams, S., Fairhead, D., Smith, R. and Ravat, D., 2008. Interpretation of magnetic data using tilt-angle derivatives. *Geophysics*, **73**(1), L1–L10. DOI: 10.1190/1.2799992.

Salem, A., Williams, S., Samson, E., Fairhead, J. D., Ravat, D. and Blakely, R. J., 2010. Sedimentary basins reconnaissance using magnetic tilt-depth method. *Exploration Geophysics*, **41**(3), 198–209. DOI: 10.1071/EG10007.

Smith, R. S., Thurston, J. B., Dai, T. F. and MacLeod, I. N., 1998. iSPI[TM]: The improved source parameter imaging method. *Geophysical Prospecting*, **46**(2), 141–51. DOI: 10.1046/j.1365-2478.1998.00084.x.

Smith, R. S., Thurston, J. B., Salem, A. and Reid, A. B., 2012. A grid implementation of the SLUTH algorithm for visualising the depth and structural index of magnetic sources. *Computers & Geosciences*, **44**, 100–8. DOI: 10.1016/j.cageo.2012.03.004.

Spector, A. and Grant, F. S., 1970. Statistical models for interpreting aeromagnetic data. *Geophysics*, **35**(2), 293–302. DOI: 10.1190/1.1440092.

Vacquier, V., Steenland, N. C., Henderson, R. G. and Zeitz, I., 1951. *Interpretation of Aeromagnetic Maps*. Geological Society of America Memoir 47. Boulder, CO: Geological Society of America.

Verduzco, B., Fairhead, J. D. and MacKenzie, C., 2004. New insights into magnetic derivatives for structural mapping. *The Leading Edge*, **23**(2), 116–19. DOI: 10.1190/1.1651454.

Vixo, D and Connard, G., 2020. Isostatic gravity inversion: A new way to model gravity data. *First Break*, **38**(5), 43–51. DOI: 10.3997/1365-2397.fb2020033.

Watts, A. B., 2001. *Isostasy and Flexure of the Lithosphere*. Cambridge: Cambridge University Press.

Werner, S., 1955. Interpretation of magnetic anomalies at sheet-like bodies. Sveriges Geologiska Undersökning, Årsbok, **43**(1949), 6.

5

Full Tensor Gradiometry

MATT LUHESHI

5.1 Introduction

In modern exploration, as discussed in Chapter 4, gravity and magnetics data are often used to help understand regional basin architecture. There are, however, many historical examples of major oil and gas fields being discovered using gravity and magnetic data alone (e.g. a number of the Middle East fields were discovered in this way). Generally, however, conventional gravity/magnetic technology is by and large a regional exploration tool and most valuable for imaging the overall fabric of basins.

Modern exploration, however, relies on high-resolution reflection seismic. Seismic allows subsurface imaging with very high resolution, enabling mapping of stratigraphic and structural features in great detail. As described in Chapter 1, explorers use increasingly sophisticated methods for understanding controls on plays and for characterising prospects. Seismic is critical for de-risking exploration, appraisal and development drilling.

Hence in modern exploration, explorers apply potential field tools when for some reason seismic of an acceptable coverage and/or quality is not available. Nowadays application of potential field methods tends to fall into one of the following general categories:

- Mapping regional basin architecture (early stages of exploration)
- Imaging in zones of poor seismic data (sub salt/basalt, structurally complex areas)
- As input to the construction of velocity models for used in seismic imaging (seismic migration)

Conventional gravity measures *one component* of the three-component gravity field (usually the vertical component – but see Section 4.4 in Chapter 4).

In contrast, however, Full Tensor Gradiometry (FTG) measures the *derivatives* of all three components in all three directions.

Airborne gravity acquisition has been well established (Chapter 4), but this suffers from resolution limitations due to the noise created by platform accelerations. Moving platforms are subject to continuous accelerations and these translate directly as noise to gravity meters. Gyroscopically stabilised platforms and the routine use of GPS have enabled reduction of this noise. However, this problem creates a significant limit on resolution when trying to measure relatively short wavelength features using conventional gravimeters deployed on aircraft or ships.

While the measurement of gravity gradients on land has a very long history, it was the development of instruments capable of use on moving platforms that has led to the explosion in the use of gradiometry in exploration. The ability to acquire FTG data using moving platforms produced a major breakthrough.

Gravity gradiometry measures changes of gravitational acceleration between two points in any direction. The gradient is simply the change divided by the distance (very small to allow accuracy) between the two points. The principle is very simple, as illustrated in Figure 5.1. A basic instrument consists of a pair of

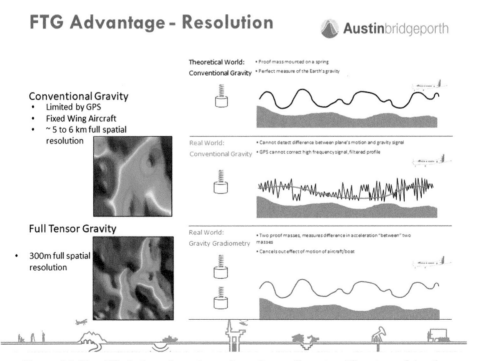

Figure 5.1 Simplified view of gravity and gravity gradiometry. (Courtesy of Austin Bridgeporth)

closely spaced gravimeters. Using two paired gravimeters enables the system to suppress noise due to platform accelerations by simple subtraction.

The great advantage of airborne and marine gradiometry acquisition systems is in their ability to reduce noise and to acquire data rapidly and accurately. Sensitivity of ~3–8 eötvös units $(10^{-9}/s^2)$ coupled to use of sufficiently close acquisition line spacing can resolve wavelengths of ~300–1,000 m.

Conventional single component airborne gravimeters are unable to reduce platform accelerations and must rely on GPS-derived acceleration corrections and filtering, which limits resolution of airborne gravity surveys to ~3–4 km typically.

This new technology is often referred to as Gravity Gradient Imaging (GGI), Gravity Gradiometry or Full Tensor Gravity (FTG). The new GGI instruments strongly suppress platform motion (turbulence or heavy sea state) by measuring differences in the gravitational field at the various sensors.

A cost-effective method of measuring FTG is a relatively recent innovation. The technology became commercially available in the early 1990s. Modern FTG data acquisition records a full 3D record of the earth's gravitational field. Its value in exploration emerged when it became possible to record such data using moving platforms – aircraft and ships.

Deployment of FTG using moving platforms has enabled two major innovations for exploration applications. It introduced a major improvement in lateral resolution as well as the ability to accurately locate the horizontal position of anomalies (because of the 3D nature of the measurements). This adds the following capabilities to conventional gravity methods:

- Radical improvement in spatial resolution – enabling very detailed structural and stratigraphic imaging
- Very accurate spatial (horizontal) location of anomalies at prospect scale
- Helps to address the problem of structural aliasing inherent in 2D seismic data interpretation
- Enables further improvements in helping to construct better velocity models for seismic imaging (seismic migration)

Figure 5.2 shows an example of the radical uplift in resolution that is achievable using FTG. The improvement in resolution and definition is clear.

As depth increases the difference in resolution between GGI/FTG and conventional gravity decreases. For target depths greater than ~3 km, the advantages of FTG decrease and it becomes questionable whether the added cost/ effort needed for FTG is appropriate. This issue is particularly important for projects in deep water.

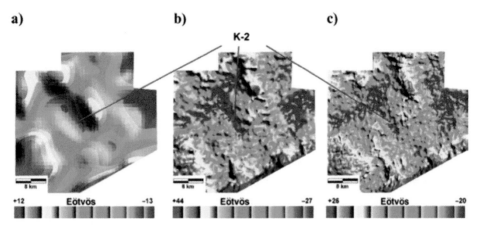

Figure 5.2 G_{zz}, the partial derivative with respect to the z-axis of the vertical force of gravity. (a) G_{zz} derived from conventional Bouguer gravity. (b) G_{zz} as measured with an FTG survey. (c) Best estimate of G_{zz} obtained by using all tensor components. (After Nabighian et al., 2005, figure 3)

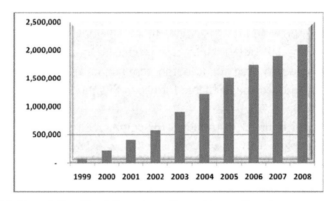

Figure 5.3 Cumulative line-kilometres of gravity gradiometer surveys (airborne and marine) conducted from 1999 to 2008. (After DiFrancesco et al., 2009)

The commercial application of FTG started in the early 1990s with steady growth in the first decade of the twentieth century (see Figure 5.3 and the historical review in the text that follows).

5.1.1 Historical Review

Measurement of gravity gradients on mobile platforms was first developed for the US Navy in the 1960s for subsea navigation. The ability to accurately locate navigation hazards was a major military application. The original military tools reportedly also included detection of nearby enemy submarines.

This military need was to be able to measure *directly* gravity gradients in 3D on a moving platform. This included the requirement to suppress significantly the effect of the accelerations due to the moving platform.

The US military funded a number of organisations to produce a suitable tool. These included Bell Aerospace (now part of Lockheed Martin), Hughes Aircraft and Massachusetts Institute of Technology (MIT). The navy eventually chose the Bell Aerospace variant, which was called the Full Tensor Gradient System or FTG. The system was eventually declassified in the early 1990s as the Cold War came to an end. Regardless of this declassification, the technology continues to be controlled by US regulators and restrictions still apply on where it may be deployed.

The technology was released for commercial trials in 1994.

Use of the instrument was later extended to onshore and airborne applications (on a C130 aircraft) by the US Air Force. The original FTG systems were bulky and not easily portable.

Bell Geospace later acquired rights to use the instruments in marine surveying and BHP-Billiton sponsored Lockheed Martin to develop an exclusive and specific airborne version (the Falcon™) which first produced good airborne data in 1999. Later the FTG was converted for airborne deployment in fixed wing aircraft by Lockheed Martin and became available for use in this mode in 2002 (DiFrancesco and Talwani, 2002).

Lockheed Martin built 12 instruments for commercial surveys. These systems were originally deployed by three companies (viz ARKeX, Bell Geospace and CGG Airborne). The first commercial application was a marine survey by Bell Geospace in the Gulf of Mexico in 1998. This was followed by commercial airborne deployment in 2002.

The Falcon™ and FTG systems have since been used in numerous airborne and marine acquisition projects. The choice of approach depends on a combination of technical objectives, cost and geography (e.g. in some locations an airborne survey would be difficult/impossible to acquire for safety reasons, such as in remote marine locations).

5.2 Theoretical Background

GGI provides a direct measurement of all three components of the spatial *second* derivatives of the earth's gravitational field $\phi(r)$. The technology gives a genuine 3D measurement of anomalies. Conventional gravimeters, however, measure just the *first* vertical derivate of $\phi(r)$, often referred to as g_z.

5.2.1 Some Basic Physics

As discussed in Chapter 4, conventional *gravimeters* normally measure just ONE component of the gravity field (often referred to as g_z, vertical gravity), as shown in Figure 5.4.

Gravity *gradiometry* is the second derivative of the scalar potential. This gives a complete description of ALL components of the gravity field and is a tensor usually expressed in matrix form (Figure 5.5; Houghton et al., 2014).

Since gravity gradient is a *second* differential of gravity (which is a first derivative of the potential field) the former has a better spectral response at higher frequencies and thus offers greater resolution (Figure 5.6), at the expense of accuracy at longer wavelengths.

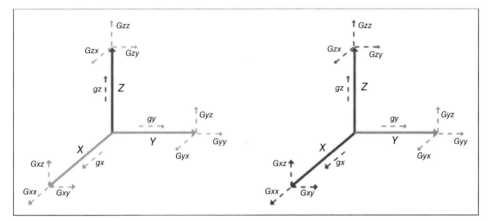

Figure 5.4 The vertical derivative of the earth's gravity potential field, g_Z, and gravity gradient measurements. (From Houghton et al., 2014)

$$G = \begin{bmatrix} \dfrac{\partial g_x}{\partial x} & \dfrac{\partial g_x}{\partial y} & \dfrac{\partial g_x}{\partial z} \\ \dfrac{\partial g_y}{\partial x} & \dfrac{\partial g_y}{\partial y} & \dfrac{\partial g_y}{\partial z} \\ \dfrac{\partial g_z}{\partial x} & \dfrac{\partial g_z}{\partial y} & \dfrac{\partial g_z}{\partial z} \end{bmatrix} = \begin{bmatrix} G_{xx} & G_{xy} & G_{xz} \\ G_{yx} & G_{yy} & G_{yz} \\ G_{zx} & G_{zy} & G_{zz} \end{bmatrix}$$

Figure 5.5 The second derivative of the earth's gravity potential field is a tensor quantity.

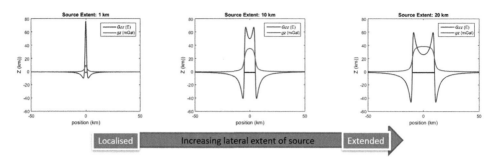

Figure 5.6 Vertical gravity (g_z) and gravity gradient (G_{zz}) signals from sources buried at 1 km depth. (Courtesy of Austin Bridgeporth)

It is important to note that although the FTG instrument measures nine components only five of these are truly independent: the remaining four components can be derived as a linear sum of the five unique measurements.

It is the ability to *measure* all nine tensors (with five independent channels) simultaneously with the possible inclusion of a 'conventional gravity' detector that leads to the noise reduction and resolution benefits behind the impact of gravity gradiometry. When conventional gravity data are also acquired in tandem with a gravity gradiometry survey, their combination (high frequencies from FTG and low frequencies from conventional gravimeters) can produce what is sometimes referred to as 'wide band gravity' (e.g. Bell Geospace's enhanced gravity system; Jorgensen et al., 2006).

5.3 Practical Application

5.3.1 Data Acquisition

Efficient acquisition of conventional gravity data requires mobile deployment on aeroplanes or marine vessels. In principle one can derive gradients from such data by simple differentiation. The problem is that this suffers from the spatial sampling of the original measurements and is subject to a bandwidth limitation.

While it is possible to use measurements from mobile conventional meters to derive gradients, such results will be subject to the same errors as the original gravity measurements that are created by random movements of the platform. Moving platforms are subject to unpredictable accelerations (primarily turbulence and wave motion). Such accelerations are registered and detected by gravimeters and appear as noise on records. This makes it difficult to de-convolve gravity changes due to subsurface effects from signals generated by extraneous platform movements.

Figure 5.7 Lockheed Martin Gravity Gradiometer System (platform, electronics and gradiometers) (DiFrancesco, 2007). (Left) FTG (Gradiometry.com). (Right black unit) FALCON™.

To mitigate this noise problem, FTG instruments (such as the Lockheed Martin FTG), use pairs of opposed accelerometers to make the gradiometers (Figure 5.1). Subtraction of the signals from two accelerometers (fixed to the same moving platform) effectively attenuates the extraneous accelerations due to platform motion. It is this engineering breakthrough that enabled this 'differencing' principle which underpins the uplift in noise reduction. This allows FTG instruments to deliver a very high signal-to-noise (S/N) ratio.

It should be noted, however, that individual accelerometers do not necessarily have the same gain; hence there is a need for calibration prior to operating a survey.

The unit of gravity gradients is eötvös (E_o),

$$1\ E_o = 10^{-9}\ s^{-2}$$

although mgal/m (or mGal/m) is also used with 1 mgal/m = 10,000 E_o.

The most commonly used commercial instruments were all constructed by Lockheed Martin. All these systems are of the rotating wheel type and consist of one or more gravity gradiometers mounted on a rotating wheel and fitted to stabilised gimbals (Figure 5.7).

There are 12 instruments available through 3 main contractors (Bell Geospace, AustinBridgeporth and CGG). The systems operated by Bell Geospace and Austin Bridgeporth are full tensor gradiometers and measure all nine components of gradiometry.

The FTG and Falcon™ have different designs (although the fundamental components are the same in that they both use paired accelerometers).

The Lockheed Martin based FTG measurement system consists of three smaller gravity gradiometers (Figure 5.7), each of which consists of four gravity accelerometers mounted on a rotating disk (Nabighian et al., 2005). The disk spin axes are mutually perpendicular, arranged in an 'umbrella' configuration. The instrument measures nine components of gradient from which five *independent components* can be constructed. The remaining gradient tensor components can be derived mathematically using symmetry.

CGG operates six Falcon™ AGG (Airborne Gravity Gradiometer) instruments. The Falcon™ version was originally commissioned by BHP Billiton as an airborne-specific instrument with maximum resolution and minimum turbulence sensitivity. It is also based on the rotating disk system developed by Lockheed Martin (formerly Bell AeroSpace). The design differs in that it consists of a *single gradiometer* disk of twice the radius, spinning about a vertical axis. It has eight accelerometers mounted round its edges (as opposed to four on each FTG disk). This version measures the *horizontal gradients* of the *horizontal components of gravity*. The vertical gradient of the vertical component of gravity is *derived by Fourier transformation*. Unlike the full FTG, Falcon is used for airborne surveys only.

The FTG unit is approximately 1 m in each dimension. The Falcon™ is commonly offered either in a small aircraft (e.g. Cessna Grand Caravan) or a helicopter. The FTG system and its electronics is presently somewhat bigger and heavier and is typically offered in a larger aircraft such as a Basler or a Twin Otter. It is also possible to deploy FTG systems in relatively small aircraft such as Cessna Caravans, although this is relatively less common.

While the two instrument types have different spectral and signal/noise characteristics (see Barnes et al., 2011 and Dransfield and Christensen, 2013), they both basically produce measurements of the gradiometry tensor. Their detailed differences need to be addressed at the level of specific survey designs.

Figure 5.8 shows an example of tensor data measured by an FTG and illustrates the geological information that may be extracted from the various tensor components. (Note that in Figure 5.8 the vertical gravity gradient is referred to as V_{zz}, etc.). The simplest component to interpret is the V_{zz} (the vertical gravity gradient), as this shows density anomalies in their correct spatial location and is most easily compared to conventional gravity.

Historical Production and Costs

Some 1,000,000 line km of gravity gradient data were acquired in the period 2002–2012. Survey sizes are highly variable, with the largest airborne FTG survey

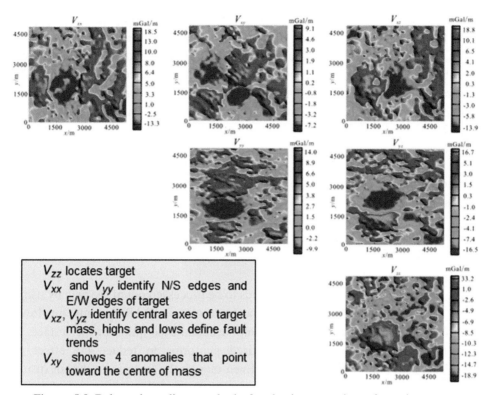

Figure 5.8 Balanced gradient methods for the interpretation of gravity tensor gradient data (from Pengyu Lu and Guoqing Ma, 2015). (Note 1 mgal/m = 10^4 E_o). V_{zz} locates the target. V_{xx} and V_{yy} identify N/S edges and E/W edges of the target. V_{xz} and V_{yz} identify central axes of the target mass; highs and lows define fault trends. V_{xy} shows four anomalies that point towards the centre of mass.

covering approximately 100,000 line km, being that acquired for Tullow in East Africa (Bell Geospace & ARKeX), and the biggest marine survey being some 50,000 km² offshore NE East Greenland (ARKeX/ION GeoVentures).

Current acquisition volumes are around 300,000 line km/year.

Acquisition costs are, of course, highly variable. Apart from the sensitivity to basic geophysical design needs, issues such as mobilisation and weather add variable costs. Historically the typical range for an FTG survey was around \$130–\$175 per line km (DiFrancesco, 2011). This compared with prices in the range \$40–\$70 per line km for conventional airborne gravity. Additionally, FTG is generally flown at closer line spacing than gravity, to achieve the higher resolution that is available in principle.

FTG is generally considerably cheaper than both 2D and 3D seismic on a \$/km or \$/sq km basis. Gradiometry, however, is 5–10 times more expensive than conventional airborne gravity, partly because of the higher price/line km and because of the closer line spacing. Hence a business case is needed for the use of FTG based on value for money relative to seismic costs and greater resolution when compared with conventional gravity. FTG delivers genuinely additional and valuable information but is most effective when integrated with seismic. As illustrated in the case studies that follow, in the *right circumstances* significant business impacts can be achieved by reducing uncertainty and risk.

Noise

Gradiometry measurements are subject to a number of sources of noise, which impose limits on sensitivity and resolution. Individual instruments have intrinsic noise levels approaching ~ 1–3 $E_o/\sqrt{Hz}$ when measured in the lab (DiFrancesco and Talwani, 2002). This allows very small variations to be measured and brings focus to the need for reducing the effect of other sources of noise (especially terrain and overburden effects).

Elevation/terrain 'noise' basically needs to be processed out of the data by using very accurate digital terrain models, and quite often LIDAR systems are flown with FTG surveys to provide a sufficiently accurate description of the surface.

Instruments are mounted on stabilised platforms to reduce the effect of turbulence and platform accelerations. Even though the instrument design (using twinned gradiometers) has an inbuilt propensity to minimise platform acceleration noise, sensitivity imbalances still leave a significant level of turbulence-induced-noise.

Post survey data processing is used to reduce this noise. Barnes and Lumley (2010) and Dransfield and Christensen (2013) give a good summary of the various sources of noise and approaches to mitigating these (see also Figure 5.9). Mitigation is through very careful survey design and to an extent in processing. The analysis of noise sources and remedies to these is a very complex subject and very much in the realm of the acquisition design and processing specialists. It is worth noting that the noise characteristics of the full FTG and Falcon™ variants of the Lockheed GGI instruments do have significant differences, which need to be taken into account when designing specific surveys (Dransfield and Christensen, 2013).

There is another form of 'noise' or uncertainty if it is desired to produce gravity data (commonly called g_z) from the gradient data, which requires an integration step. This has uncertainties at the longer wavelengths. There are various strategies to mitigate this (including operating a gravimeter in the same aircraft alongside an FTG or merging with external control data). Another approach that has been deployed more recently is to use the gravity module assembly (GMA) systems as described in Barnes (2018), where a conventional gravity channel that is part of the

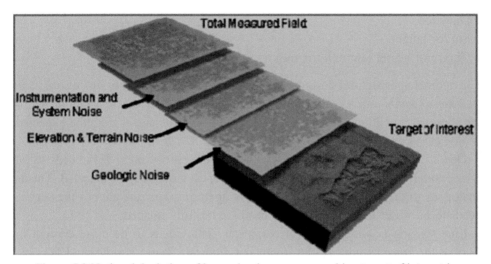

Figure 5.9 Notional depiction of layered noise sources masking target of interest in a geological survey. Noise includes contributions from sensors and systems, topology, and underlying geological variations. (After figure 2 in DiFrancesco et al., 2009)

Lockheed Martin FTG system is activated and data acquired at the same time as a gradiometry survey.

Survey Design

A good description of the issues that need to be considered when designing an FTG survey is given in Dransfield and Christensen (2013). The main requirements from good design are to deliver accurate data with as high a S/N ratio as possible/ affordable. Note that there are some differences between AGG and FTG instruments and hence survey designs. What follows focuses on FTG, as this is more commonly used in oil and gas exploration.

Raw FTG measured data will consist of the desired subsurface geological signal in combination with 'noise' that is due to terrain variations, overburden effects and intrinsic instrument noise. Survey design and careful operational planning are the most important tools in use to ensure sufficient sampling to allow effective noise reduction and resolution enhancement in processing. The survey planning will also specify the acceptable range of operational parameters (turbulence, etc.).

Design needs to consider the following elements:

- Geological target
- Areal coverage
- Line spacing

- Direction
- Survey height
- Choice of vessel (aircraft, marine)

The optimal approach for designing a survey is to build feasibility models that can take these factors into account. Feasibility models can help to assess the impact of the various acquisition parameters on detectability of the signal from the expected geological target.

One critical parameter choice is the size of the survey area. This needs to be wide enough to ensure that signals from the deepest target may be detected. This is analogous to the familiar seismic migration aperture issue: the area of the survey needs to be wider than the zone that needs to be fully imaged.

Line spacing is another critical parameter. This needs to be close enough to resolve the wavelengths at the depths of interest. Typical airborne FTG line spacings are in the range of 300 m–2,000 m. Unlike in seismic migration, oversampling is not necessary for FTG. The 3D tensor nature of the FTG records enables accurate spatial interpolation. Clearly this can have a profound impact on acquisition cost.

The need for maximising the recording of what are very weak signals and minimising platform noise imposes restrictions on survey operations. Typically, this means flying as low and slow as possible. Typical survey heights for airborne surveys are between 80 and 150 m and at speeds of 50–60 m/s). This has obvious consequences on acceptable weather conditions and consequently on costs.

5.3.2 Data Processing

Gradiometry data require considerably more processing effort than conventional gravity measurements. The need for co-processing five independent measures of gravity gradient, together with an independently measured *gravity* signal, adds considerable complexity. The benefit, though, is that the constraints imposed by these additional data provide enhancements in resolution and noise rejection.

A detailed description of the processing of gravity gradiometry data is given in Jorgensen et al. (2006), Saad (2006) and Barnes and Lumley (2011). The processing approaches are based on recognition of the fact that all the components of the gradient tensor originate from the same mass distribution. Hence all components of the gravity field need to be processed jointly to satisfy this geological constraint.

Barnes and Lumley (2011) describe two general approaches to data processing: (1) a Fourier transform approach and (b) a Common Equivalent Source method.

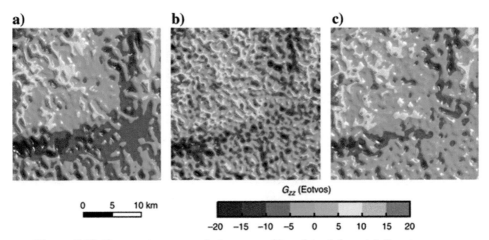

Figure 5.10 G_{zz} terrain-corrected, low-pass filtered to 1 km. (a) Exact answer. (b) Fourier processed. (c) Equivalent Source processed. All data are mean de-trended and plotted with the same colour scheme. (After Barnes and Lumley, 2011, figure 14)

The authors argue that the latter method is generally more accurate and more effective at reducing noise (by a factor around 2.4; see the example in Figure 5.10), although it is computationally much more expensive. Development of efficient numerical algorithms has made the 'equivalent source method' the routine approach to data processing with at least this one contractor.

5.3.3 Interpretation

Interpretation of *conventional gravity* data is relatively straightforward, as it basically involves just the z (g_z, vertical) component of the field. A gradiometry data set is fully described by a nine (3×3) matrix (of which five are independent). The different gradient combinations are illustrated in Figure 5.11; please note that this figure uses 'T' as the symbol for gravity gradient; e.g., T_{zz} is equivalent to G_{zz} in this text). The easiest component to understand is T_{zz} (the vertical component in Figure 5.11) which shows an anomaly that directly overlies the model prism target. The other components provide information related to the edges and lateral location of the source. In fact, it is availability of the horizontal tensor components that gives FTG such an advantage, as these are used to accurately locate the edges of the source.

The importance of the additional information provided by the tensor gradients is illustrated in a very elegant model example given by Barnes (2012). The following figures (Figures 5.12 and 5.13) show a model consisting of three anomalous masses, which are then 'over flown' by three survey lines (densely sampled in the inline direction and measured every 2 km in the cross line direction). Line

(a)   (b)

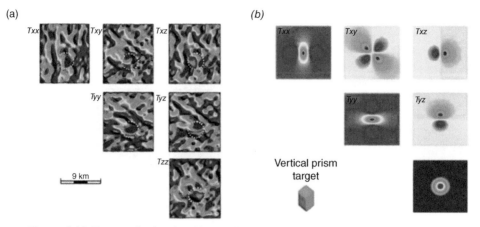

Figure 5.11 Tensor display for (a) an Air-FTG™ survey acquired over a known salt caprock feature onshore USA and (b) a theoretical response for an idealised target. The white polygon helps to locate the response of the caprock in each component. (After Murphy 2004, figure 2. © Commonwealth of Australia [Geoscience Australia] 2020. This product is released under the Creative Commons Attribution 4.0 International License. http://creativecommons.org/licenses/by/4.0/legalcode)

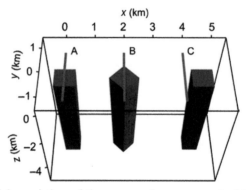

Figure 5.12 Model consisting of three anomalous masses (red blocks) and three survey lines (blue lines). (After Barnes, 2012)

A crosses directly over the block as does Line B. Line C is offset 500 m from centre of the third block. The experiment is assumed to be noise free. Barnes (2012) shows the results of an inversion for density distribution using two sets of assumptions (Figure 5.13): the first (a) assuming that only the vertical G_{zz} component is available and the second (b) where all the tensor elements are included.

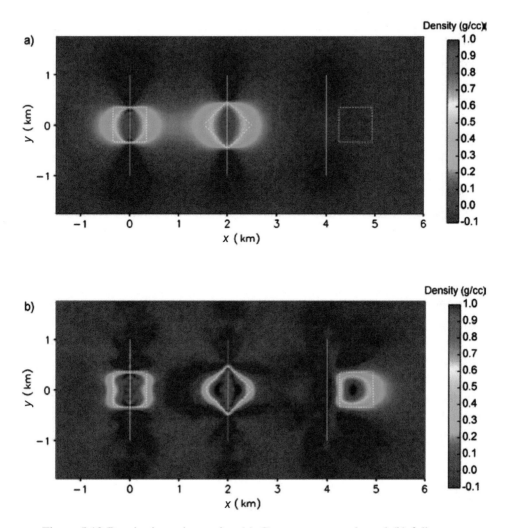

Figure 5.13 Density inversions using (a) G_{zz} component only and (b) full tensor data. Yellow lines represent the three survey lines, dashed outlines show the actual source locations (Barnes, 2012).

Figure 5.13 a shows the result when only G_{zz} is used and, as expected, this shows density anomalies beneath lines A and C which are vertically below the relevant acquisition lines. For line C, though, there is only a very weak signal, as here the source is offset from the line.

Figure 5.13b shows the result using all tensor components and this shows a very accurate reconstruction of all three density anomalies in terms of both shape and location. Even in the case of the third prism (C), which is offset from the acquisition line C, an accurate density image *and* location is recovered.

This simple model illustrates what is possibly the most important benefit of using full tensor gradiometry. The ability to image accurately the location and orientation of edges of sources (lateral density changes) accurately is of major benefit in structural correlation. For an example like case (C) conventional gravity and 2D seismic would show an incorrect location. 3D seismic data would be needed to give a result similar to FTG (3D seismic, however, would also give an accurate depth in addition).

For a real example such a result would clearly depend on a number of environmental parameters as well as being sensitive to noise levels. It does make the point, however, that the additional tensor information can potentially transform an interpreter's ability to map structural correlations accurately when only 2D seismic and/or conventional gravity data are available (with some care).

A real data example (Figure 5.14) in Barnes (2012) examines the incremental value of finer sampling. The horizontal tensors clearly provide gradient information that can be used in interpolation and accurate location of anomalies. Figure 5.14a shows the vertical component G_{zz} from a real FTG survey acquired at 150 m line spacing. Figure 5.14b shows a reprocessed version of subsampled data using 5 km line spacing. It is clear that both attempts show the same gross features in the same locations, with the full data set giving a much finer resolution, as expected.

This offers the possibility of designing surveys in two stages: using relatively wide line spacings for a more 'regional' picture, followed by an infill programme to investigate particular anomalies in more detail. This would be an advantage in terms of cost efficiency. A large area can be 'screened' initially using wide line

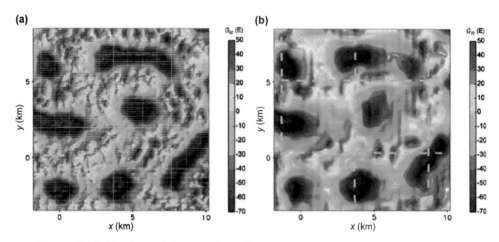

Figure 5.14 (a) 12 × 12 km section of survey showing G_{zz} deduced from the processing of full tensor data over closely spaced survey lines. Faint grey lines show survey flight lines. (b) G_{zz} deduced from processing only the data from flight lines separated by 5 km (shown as grey dashed lines).

spacings. Higher resolution surveys can then be acquired to infill in the required focus areas.

The higher resolution and greater fidelity (in terms of horizontal location of anomalies) relies on the availability of the additional tensor information. Although it can be challenging to interpret the full tensor suite (see, e.g., Saad, 2006) much of the value can be extracted from the G_{zz} component, which is more intuitive to interpret. G_{zz} derived from the fully integrated processing of a full tensor data set, necessarily includes the effect of the remaining tensors and hence gives a reliable image of anomalies in its own right.

Forward modeling and inversion analysis is also used to test geological interpretations, in a way that is similar to the integration of conventional gravity surveys.

5.4 Case Studies

Gravity gradiometry imaging relies fundamentally on lateral density contrasts and is identical to 'conventional' gravity surveying in this respect. This implementation of the gravity methods has two basic advantages due to improved signal-to-noise characteristics and to the fact that FTG instruments measure all three components of the gradient of the gravity field simultaneously. The case studies described in this section illustrate these two points. The exploration applications fall into one or a combination of the following categories:

- Direct higher resolution imaging of structural fabric in areas where seismic is compromised
- Construction of better constrained seismic velocity fields in service of improving seismic imaging (2D and 3D)
- Mitigating the problem of aliasing when correlating structural features using 2D seismic (i.e. absent 3D seismic)
- Direct detection of reservoir lithologies
- Direct imaging of prospect scale features

In all applications, the exploration impact is maximised when a variety of technologies are integrated intelligently. The relative contribution of any given technology to any specific exploration problem, is of course case specific.

The following cases illustrate some of these points.

5.4.1 Structural Correlation (Using 2D Seismic, FTG and AGG Data)

There are a number of published cases illustrating the use of tensor gradiometry to improve fault correlation in the absence of 3D seismic. Figure 5.15 shows one such

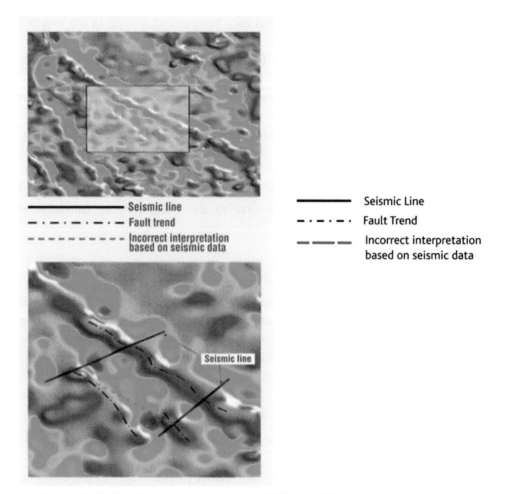

Figure 5.15 Correlating structure using FTG and 2D seismic. (Courtesy of IONGEO, pers. comm Phil Houghton)

example given by Houghton et al. (2014) and shows the aliasing risk inherent in using 2D seismic data without any additional information. A seismic interpreter has picked the dashed line as the trend of a fault seen on the two seismic lines. Clearly, however, the G_{zz} anomaly shows a different correlation which could have a profound impact on the geological interpretation. The G_{zz} map gives a much more accurate measured location for the fault. This addresses a very common problem with 2D seismic interpretation. Prior to the availability of FTG data, the only way to resolve such problems would be to acquire a much more expensive 3D seismic survey.

Of course (for this and all other similar situations) it will be necessary to check this by using inversion and/or forward modelling to establish that the observations

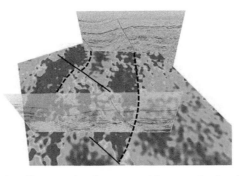

Figure 5.16 Example of interpreting between wide spaced seismic lines with AGG data to guide a structural interpretation. Blue faults on seismic are joined with confidence (black dashed trace) and transfer faults readily identified. (After figure 7 from Moore et al., 2012)

on the G_{zz} map are indeed from the correct depth (i.e. the map is imaging the fabric that is of interest).

This is an example of probably the most valuable application of FTG data. The ability to confidently follow the trace of structural anomalies between widely spaced 2D seismic lines adds a major degree of confidence for structural interpretation at high resolution (subject to depth, of course).

Figure 5.16 shows another example of such an application published by Moore et al. (2012), this time using a Falcon™ AGG instrument. This is a clear example of interpolating faults seen on regional 2D seismic lines using gradiometry data.

5.4.2 Gabon Salt and Carbonate Imaging

Gravity data are commonly used to distinguish lithologies that have a high-density contrast, for example, salt or carbonates deposited within a siliciclastic depositional framework. Gravity gradiometry also has this capability, clearly.

A good example is a case study reported by Davies and Martin (2010). They describe a joint CGG/Veritas/ARKeX project in which they acquired marine FTG data over an area of 9,000 km² and then integrated this with a regional 2D seismic grid. The study was located in the Gabon Atlantic Margin in deep water and was able to image usable FTG anomalies down to around 8 km depth.

The geology of the province included carbonates and salt, thus providing some large density contrasts.

Figure 5.17a shows G_{zz} with a 20 km high-pass filter applied showing improved delineation of carbonates (in red) and salt (in blue). Notable is the imaging of the carbonate structures which is the result of the added resolution provided by FTG. Figure 5.17b shows G_{zz} as computed from a *conventional marine gravity* survey,

(a) Gravity Gradiometry - G_{zz} with   b) Computed vertical gradient from marine
    20 km high-pass filter   Gravity (G_z) data with 20 km high-pass filter
                                               applied to existing marine gravity data

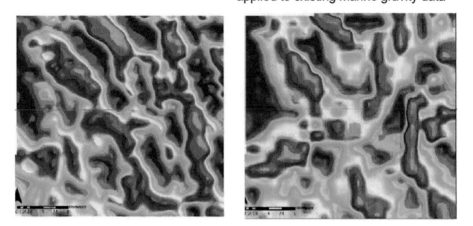

Figure 5.17 Comparison of gravity gradient from FTG survey (a) and gradient derived from conventional marine gravity data (b), offshore Gabon. (a) Gravity gradiometry - Gzz with 20 km high-pass filter. (b) Computed vertical gradient from marine gravity (G_z) data with 20 km high-pass filter applied to existing marine gravity data. (After Davies and Martin, 2010)

also filtered using a 20 km filter. There are two clear differences here: (a) the FTG example has higher resolution and (b) there is a clear difference in the morphology of the anomalies. Comparison of these two anomaly maps with 2D seismic shows that the FTG image gives a closer tie to the seismic.

5.4.3 Sub-Salt Imaging: Resolving the K-2 Salt Structure in the Gulf of Mexico

This example shows how the inversion of FTG data to produce an accurate depth model of the base of a salt feature was used to calibrate a 3D seismic image. The K-2 field lies some 175 miles SW of New Orleans in the Gulf of Mexico (Figure 5.18). K-2 is operated by Anadarko Petroleum which holds a 41.8% working interest in the field. The field lies in Green Canyon Block 562 in the deep water of the Gulf of Mexico.

The reservoir is of Miocene age and the trap is a three-way dip structure with a salt top seal. The original reserves were estimated to be around 100 mmbbls of oil and the top reservoir (base salt) is at an average depth of some 25,000 feet subsea (water depth here is 4,326 feet).

A key uncertainty was the extent of the reservoir up dip beneath the thick salt structure. This created a major problem for the appraisal programme. The extent of the sands up dip beneath the thick salt overhang presented a major appraisal

Figure 5.18 Location of K-2 field, Gulf of Mexico.

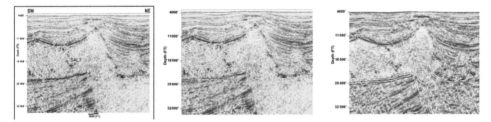

Figure 5.19 Pre-stack depth migration profile along a line through the K-2 field. (Left) Kirchhoff migration, showing poorly imaged area updip of the field. (Centre) Kirchhoff migration with base of salt horizon in yellow, as determined by FTG inversion. This shows a symmetric salt keel. (Right) Wave-equation pre-stack depth migration which also shows the presence of a salt keel, yellow horizon shows the FTG inversion result. (After figure 8 from O'Brien et al., 2005)

challenge. The image in Figure 5.19 shows a line from a pre-stack depth migrated 3D seismic survey using Kirchhoff migration. The uncertainty in the termination of the reservoir against the overlying salt is clear.

The Anadarko partnership was faced with two stark choices in managing this problem: either (a) drill an up dip well to test reservoir extent or (b) potentially leave a substantial amount of reserves undeveloped. The geophysical challenge was clear and conventionally the 'standard' approach would be to reprocess the 3D with newer/better imaging technology. As well as this, Anadarko decided to also acquire a gravity gradiometry survey for two reasons: (1) to help with defining the velocity structure for a new depth migration and (2) to get an independent estimate of the base of salt morphology.

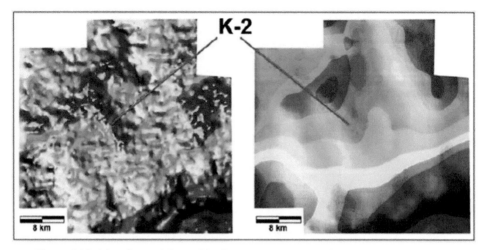

Figure 5.20 After figure 3 from O'Brien et al. (2005). (Left) G_{zz} component of the gravity gradient field over K-2. (Right) Conventional 3D free air gravity over the same area. Note the superior definition of the gravity gradient data.

Imaging a salt/soft sediment interface using FTG is optimal from a density contrast perspective; however, the 25,000 feet depth of the base of salt was a significant challenge in terms of resolution.

An image of the G_{zz} data from the FTG survey is shown in Figure 5.20, which highlights the superior resolution of the FTG data.

Prior to the reprocessing and joint FTG inversion, there were three possible interpretations of the base of salt, namely (a) a flat event across the area of poor imaging, (b) a broad salt 'keel' and (c) a shallower base salt (see Figure 5.19).

In Figure 5.19 the first image is for the original depth migrated section showing the uncertainty in the morphology of the base of salt. The second image adds the depth of the base of salt as estimated from the inversion of the FTG data and the final image shows the updated depth migration. The last version shows that the revised migration is in excellent agreement with the FTG result and hence gives confidence in the 'keel' interpretation of the base of salt.

This result had a profound implication on the field development plan. Without this the operators would have been faced with a stark choice of either (a) drilling an additional up dip well or (b) potentially leaving some pay undeveloped. The integration of FTG data with 3D seismic greatly reduced the uncertainty and hence a significant amount of cost as well as enabling a more accurate estimate of field reserves.

5.4.4 Sub-Salt Imaging in the Red Sea

This is an example of improving seismic imaging below layered evaporites through integration of FTG, Controlled Source Electromagnetic (CSEM) and 3D seismic.

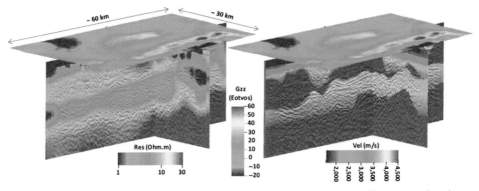

Figure 5.21 Seismic depth volume along a central inline and crossline co-rendered with (a) 3D MT resistivity inversion and (b) seismic depth velocity. The vertical gravity gradient distribution (G_{zz}) is shown in map view on top of the cross sections. The seismic volume is migrated to a depth of 9 km. (After figure 5 from Colombo and McNeice, 2013)

The effect of seismic noise generated by layered evaporites is notorious in the Gulf of Suez and the Red Sea and presents an extremely difficult seismic imaging problem. The Tertiary section of the Red Sea, for example, is characterised by thick evaporites interbedded with clastics. This creates a very difficult seismic imaging problem because of the dominance of strong interbed multiples. Colombo and McNeice (2013) describe a novel integration of multiple geophysical techniques to try and overcome this historically very difficult imaging problem.

The exploration target is in Tertiary sediments deposited over deep basement structures and the goal was to image the potential reservoirs beneath the evaporite layers.

Aramco (Colombo and McNeice, 2013) decided to acquired Wide Azimuth (WAZ) 3D seismic as well as coincident FTG, CSEM and magnetotelluric (MT). These four data sets were then combined to build more accurate velocity models for 3D seismic migration as well as building integrated inversion workflows that honoured all the geophysical datasets. This enabled them to build much more detailed sub-salt (layered evaporite) structural models (Figure 5.21).

In this example, the main role of the FTG element was to give an accurate image of the basement structures as well as assisting with building a more accurate velocity model for improved seismic migration. This example is notable for the integration of four geophysical methods in an attempt to solve a hitherto very difficult seismic imaging problem.

5.4.5 Prospect Definition in East Africa

FTG/AGG have been used very successfully for imaging prospect scale features in East African Rift (EAR) basins. A number of exploration companies have acquired airborne gradiometry data, to guide both the acquisition and interpretation of 2D seismic.

These include Tullow, Total, Taipan Resources, Tower Resources, Lundin and Simba Energy. This is testimony to its effectiveness for exploring in onshore areas where use of 3D seismic can be prohibitively expensive (typically, 3D seismic costs are around 10 times as much as AGG for an equivalent area). The EAR is a classic example of the synergy between AGG and 2D seismic.

In 2009 Bell Geospace acquired an FTG survey for Tullow to explore their licence covering the Albertine Graben of the EAR (see Figure 5.22). This survey

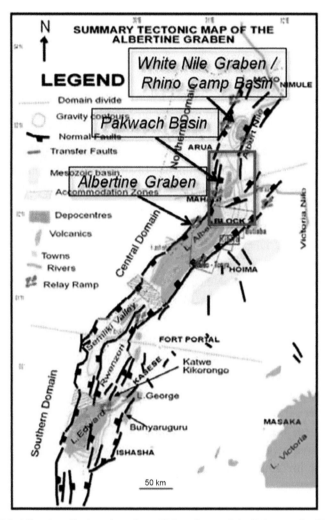

Figure 5.22 Albertine Graben location. (After figure 2 in Price et al., 2013)

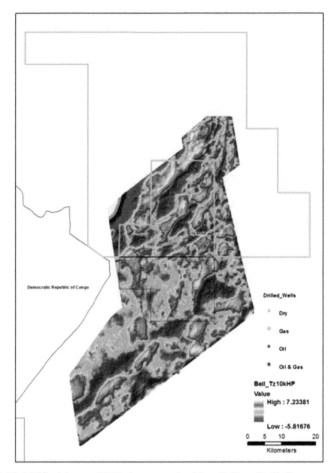

Figure 5.23 2009 airborne FTG data acquired by Tullow/Bell Geospace – 10 km high-pass of vertical Bouguer gravity with drilled wells indicated. (After figure 3 in Price et al., 2013)

was used to infill the structural interpretation from the regional 2D seismic lines. It was successful in identifying a number of structural highs that correlated very well with closures that proved to contain hydrocarbons (Figure 5.23).

In 2012 this survey was extended by Total into their licences to the north of the Tullow acreage (see Figure 5.24). This extension was acquired using the Falcon AGG system operated by Fugro. The new acquisition used different parameters (Price et al., 2013) in order to try and optimise a number of features (e.g. the orientation with respect to regional strike).

Figure 5.25 is a merge of the Tullow and Fugro surveys and clearly shows good continuity between the two vintages of data in spite of the different acquisition parameters and systems used in the 2009 and 2012 surveys.

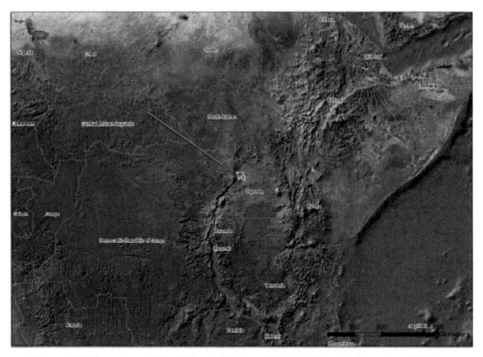

Figure 5.24 Location of Total blocks EA1 and EA-1a, Uganda (location of Falcon AGG survey outlined in blue right-hand image). (After figure 1 in Price et al., 2013)

Qualitative and quantitative interpretation (including inversion integrating with the magnetic data; see Figure 5.26) was successful in extending the detailed structural interpretation between the regional 2D seismic lines. This success was achieved primarily because of the relatively shallow target depths (2–3 km) and the high-density contrasts between the basement and sediments. Since the target plays have a close association with basement highs this meant that these FTG/AGG programmes were very successful at locating traps.

5.4.6 Integration of 2D Seismic, Gravity Gradiometry and Magnetic Data on a Passive Margin: NE Greenland

Jackson et al. (2013) describe a classic application of the integration of high-resolution FTG with long-offset 2D seismic and magnetic data to help understand the main structural elements in a frontier passive margin basin. The data set consisted of three phases of long offset 2D seismic acquired between 2008 and 2012 and 50,000 km^2 of airborne FTG (the largest offshore

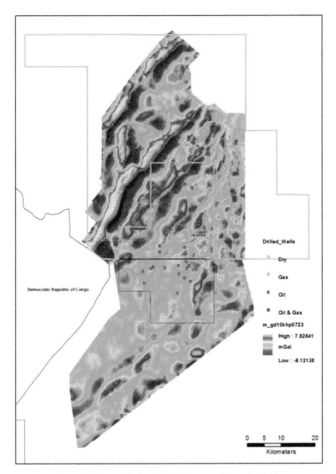

Figure 5.25 Merged AGG data 10 km High-Pass of vertical Bouguer gravity. The merge line appears as white. (After Price et al., 2013, figure 4)

FTG survey at the time). The goal was to map the distribution of igneous/ volcanic elements, geometry of salt sub-basins and fault correlations both in the basement and the overlying sediments. The setting is shown in Figure 5.27.

One example from Area B in Figure 5.27 shows the impact of FTG data on the imaging. Figure 5.28 shows seismic line B1 with an overlay of G_{zz} from the FTG survey, which shows clear evidence of two major salt features that are poorly imaged on the seismic.

Figure 5.29B shows that many of the smaller salt features imaged on the seismic actually appear as part of larger more contiguous salt bodies. Figure 5.29C shows that some of these salt features are actually salt walls extending over 60 km in length.

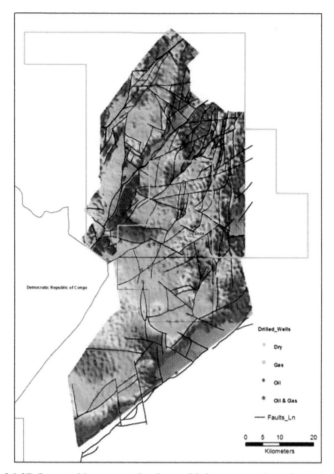

Figure 5.26 3D Inverted basement depth combining magnetic and gravity gradient tensor components. (After figure 5 from Price et al., 2013)

This example illustrates the power of FTG data to infill to high-resolution interpretations from a sparse 2D regional seismic survey. In this instance gradiometry provides critical data on the extent and morphology of salt structures that are likely to have a profound impact on the prospectivity in the area. This is at a cost that is considerably lower than that of a closely spaced seismic survey over the same area.

5.4.7 Sub-basalt Imaging: Faroe–Shetland Basin Example

Another classical application of potential fields (gravity, magnetic and EM fields) is imaging of geology beneath basalt. One notable example is the study of an area of the Faroe–Shetland Basin using a number of FTG surveys (proprietary and

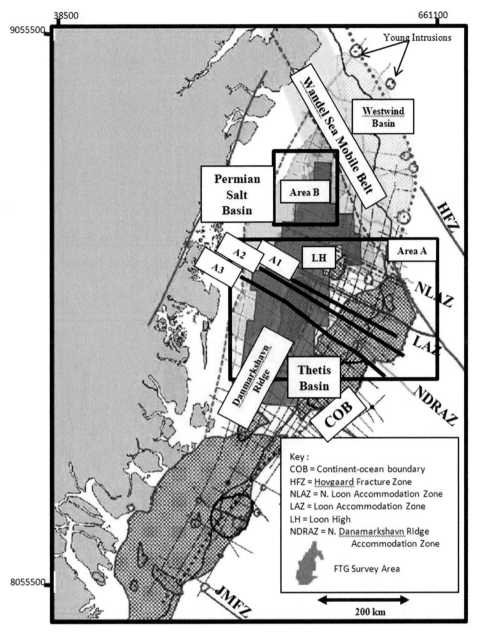

Figure 5.27 Tectonic elements of the NE Greenland passive margin. (From Jackson et al., 2013, figure 1)

multiclient) using Bell Geospace's Marine-FTG system (Marine-FTG®; Murphy et al., 2005).

Extensive parts of the Faroe–Shetland Basin are known to be affected by extrusives and very large areas of flood basalts. The basins are known to have

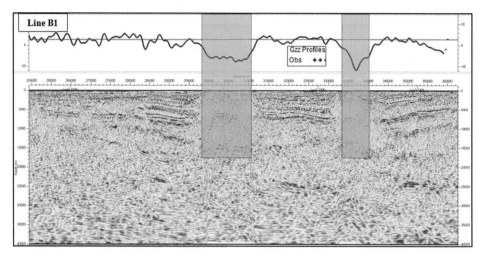

Figure 5.28 Seismic line B1 and the observed G_{zz} profile. (After Jackson et al., 2013, figure 5)

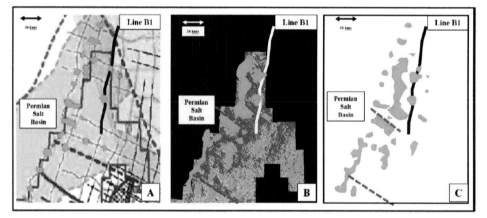

Figure 5.29 Combining regional seismic 2D with high-resolution gravity gradiometry to improve definition of salt and related structural trends. Salt imaging (A) from regional seismic, (B) & (C) using gradiometry data. (After Jackson et al., 2013, figure 6)

working petroleum systems (with the Foinaven and Schiehallion fields being the first proven giant oil fields discovered by BP in the early 1990s). Regional analysis and results from some wildcats indicate that the petroleum systems probably extend beneath the Tertiary basalts.

The flood basalts thicken towards the west into Faroese waters and create a formidable challenge for seismic imaging. Hence there have been a number of attempts to use both potential fields and very long offset seismic acquisition to try and penetrate the basalt cover.

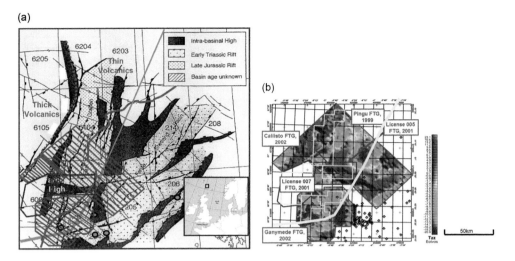

Figure 5.30 Marine FTG Survey: Faroe–Shetland Basin. (From Murphy et al., 2005) (a) Location map of Faroe–Shetland Basin area. FTG surveys shown in red (speculative) and blue (proprietary), basalt front in solid magenta. Basalt thickens NW of dotted magenta line. (b) Merged 3D-FTG Tzz component for all surveys. Drilled wells shown in black.

A number of FTG surveys were acquired by Bell Geospace using their Marine-FTG® system between 1999 and 2002 (Figure 5.30A).

Figure 5.30b shows the full unfiltered vertical gravity gradient (NB the figure uses T_{zz} rather than G_{zz}), which illustrates the main regional fabric and is affected by density contrasts at all depths.

Figure 5.31 displays a spectral analysis of Tzz and shows some interesting variations in the orientation of the structural fabric with depth. The short-wavelength (3–10 km) panel shows the shallow geology including an interpreted image of a channel (indicated in the figure). The intermediate wavelength (10–20 km) image (B), shows anomaly terminations that image the edge of the thick basalt (as independently detected by the seismic). The longest wavelength slice (>10 km) shows the trend of the regional highs which are critical for prospect de-risking. These regional highs form the sites of hydrocarbon migration foci and hence are areas of low charge risk.

This example is interesting in that it illustrates the ability of FTG to show high-resolution images of shallow stratigraphy (near surface channels), as well as the ability to image structure beneath significant thicknesses of basalt. Both results are valuable from both an exploration and imaging perspective. Direct imaging of sand-filled channels could be used to identify leads under the right circumstances. Moreover, sub-basalt images supporting the existence of thick sediments beneath basalt are critical for verifying the extension of basins beneath the basalt. These are

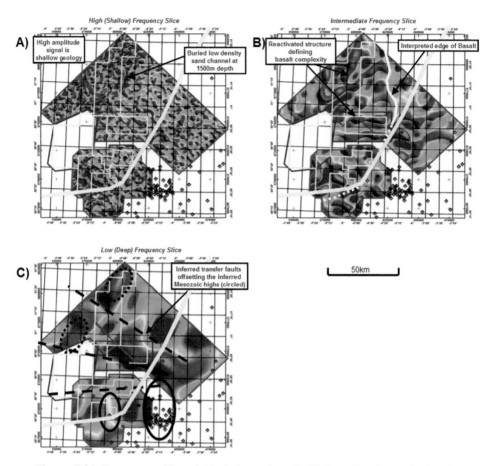

Figure 5.31 Frequency filtered T_{zz} information. Drilled wells shown in black. (A) shows short wavelenths (3–10 km), (B) shows medium wavelengths (10–20 km) and (C) shows the longest wavelengths (>10km). (After Murphy et al., 2005, figure 3)

likely to contain potentially new petroleum systems in a domain that is as yet mostly inaccessible to good quality seismic.

5.4.8 Discussion

In terms of applications FTG represents a major extension to conventional gravity. Improved sensitivity and resolution give obvious advantages. Perhaps the most profound impact comes from the tool's ability to locate anomalies accurately in the horizontal plane.

In common with conventional gravity, FTG is dependent on lateral density contrasts, wavelength and depth to the features of interest. Gradiometry depends on mass distribution and as such suffers from the same non-uniqueness issues as conventional gravity. Successful surveys, however, have been conducted in a wide range of geological environments and depths (as illustrated by the case studies described earlier).

Gradiometry has added significant value for the following categories of exploration problem:

- Clastic/salt interface systems
- Clastic/carbonate systems
- Prospect definition at relatively shallow depth (typically <2 km subsurface)
- Sub-basalt imaging
- Sub-salt imaging
- Improving seismic imaging sub-salt (and in principle sub-basalt)

Arguably the most significant uplift in exploration value is when FTG surveys are acquired as a complement to relatively sparse 2D. In some senses this raises the use of 2D seismic to '2.5 D' – the result is not as complete in an imaging sense as 3D, but for some applications is a very good interim stage (the application of FTG in East Africa is a clear case in point). Interpolation of 2D seismic images using gradiometry in some circumstances can get very close to delivering a level of prospect risk estimation that is acceptable and cost effective.

One significant challenge for gradiometry is in marine applications in deep water. Here depth to target will have an impact on sensitivity and resolution.

The complex interplay of sensitivity and noise is such that it is difficult to use intuitive rules of thumb as a guide for when gradiometry surveys could add significant value. That is why it is always advisable to conduct some feasibility studies ahead of survey design (akin to seismic survey design in some ways).

Table 5.1 shows the circumstances for which FTG can have a cost-effective impact on E&P activities.

- Green – implies a reasonable chance of success
- Orange – implies uncertainty/probably case dependent
- Red – implies successful application will probably be a challenge

In all cases, though, the devil will be in the detail. For example, water depth could be too severe for relatively deep targets below mud-line, but feasible for relatively shallow targets.

The definitions are necessarily vague as the problems will be basin specific and potential usefulness of the technology will have to be assessed locally. Nevertheless, the table gives a first pass indication of the potential of the tool in various environments and stages of the oil and gas value chain.

The significant point here is that gradiometry has the potential (proven in some cases) of assisting with exploration and appraisal well de-risking (again in the right circumstances). This has major economic benefits especially in difficult terrains.

Table 5.1 *Use of gravity gradiometry in the E&P life cycle*

	Basin definition	Exploration prospects	Appraisal	Development	Production
Onshore – low relief					
Onshore severe relief					
Shallow marine					
Deep marine					

Green indicates gradiometry is likely to be applicable in this setting; orange that usefulness is uncertain or probably case dependent, while red implies that applicability is likely to be challenging.

With careful feasibility analysis and survey design, it is clear that FTG is able to reduce uncertainty and enable better definition of risk at a number of E&P stages. The ability to do this at costs that are substantially less than equivalent 3D seismic surveys has made this a powerful addition to the explorer's tool kit.

Finally, perhaps the most surprising inclusion is the idea of using FTG in the production phase of a field (essentially for reservoir monitoring). This has been demonstrated for simple noise free models – intuitively for reservoirs with high enough porosity, replacing hydrocarbons (most likely gas) with water through production could give a detectable signal. The concept has been demonstrated using conventional gravity measurements for shallow gas storage systems. The idea of time lapse FTG for production monitoring is still at a conceptual stage.

5.5 Future Trends

Although all GGI/FTG surveys acquired to date use Lockheed Martin metres, there are several alternative systems under development. Current and future developments are focused on improving intrinsic sensor performance (sensitivity), platform noise reduction (better isolation of instruments and compensation techniques) and combining instrument types (e.g. enhanced FTG [eFTG] for improved resolution). There are a number of ongoing developments (e.g. see DiFrancesco, 2007 for descriptions). Developments include

- **Electrostatic Gravity Gradiometer.** This is the gravity gradiometer deployed on the European Space Agency's GOCE mission.
- **eFTG.** An extension of the Lockheed Martin FTG that also records scalar gravity (see later).

5.5.1 Enhanced FTG

Gradiometry provides a higher resolution than scalar gravity when measuring the gravity field from a vessel or aircraft and thus provides considerable additional detail. However, the measurement of the gravity field using gradiometers alone does have the shortcoming of not measuring the longer wavelengths, which are desirable for the sensing of 'deeper' geology. Theoretically it is possible to integrate the gradiometry data to derive the long wavelengths; but this process introduces no new information.

One option for addressing this problem is to merge gradiometry data with existing measured regional gravity (Dransfield, 2010). Integration of measured regional gravity does result in improved imaging.

A better approach is to measure the gravity field in parallel with the FTG acquisition (Barnes, 2018). The Lockheed Martin FTG instrument may be accompanied by a conventional gravimeter called a gravity module assembly (GMA). The combination of these two units on the same platform allows for greatly enhanced S/N ratios (Barnes 2018). The combination of gradiometry and scalar gravity provides improved signal/noise ratios over greater bandwidth.

The improvement to S/N ratio provides a means to resolve more subtle density contrasts. As long as the Nyquist sampling requirement is met with the survey design, wider line spacings could yield the same S/N performance compared to a tighter line spaced survey acquired with a higher noise instrument. Hence the gain in S/N from an eFTG system could either be exploited to improve definition of anomalies (i.e. create a sharper image) or it could be used to relax the line spacing compared to a higher noise system without affecting the detectability.

If the survey objectives are purely regional and no detail is required, then an airborne gravity survey should be considered on budget grounds. If, however, the technical objectives demand fullest bandwidth signal then a combination of scalar gravity and gradiometry should be the choice. Feasibility modelling is able to assist with the choices. Which strategy is used could be assessed from detailed feasibility modelling.

Barnes (2018) describes an example of an eFTG. This example shows the benefits of combining gradiometry with GMA data to provide greater accuracy and resolution.

5.5.2 Reservoir Monitoring and Carbon Capture and Storage

Production from or injection into reservoirs will alter the bulk rock density of the host lithology. When hydrocarbons (oil and/or gas) are produced the product is displaced with brine. If gas is injected, then the stored liquid (usually oil) is displaced, creating a potentially significant reduction in bulk density.

Since gravimetry depends on changes in density then it follows that it is theoretically possible to detect changes in fluid saturation using time lapse gravity measurements. A number of practical examples exist of reservoir production/ injection being monitored using gravity measurements. Although this is a relatively rare application, the following two examples present a persuasive technical case for considering gravity (and gravity gradients) as a tool for monitoring changes in fluid saturation in reservoirs.

4D seismic is routinely used for monitoring and is by far the preferred technology used by oil companies. This is due to the high resolution and spatial accuracy of repeat 3D seismic surveys. The major problem with using 4D seismic is the cost of these programmes. Methods based on gravity or gravity gradient measurements are considerably cheaper if the required sensitivity can be achieved (a signal amplitude of approximately a few to 100 μGals). The ambiguity inherent in the gravity method of distinguishing depth versus density variation would be less of an issue for reservoirs that are already well imaged and described using conventional 3D seismic. Thus, in principle it may be possible to integrate repeat gravity surveys into a reservoir monitoring programme. This would be complementary to conventional 4D seismic surveys and hopefully reduce the number of time-lapse seismic surveys needed for accurate monitoring.

This need for sensitive geophysical monitoring of fluid movement in reservoirs is particularly important for the emerging need for effective and efficient methods to monitor disposal of CO_2 in subsurface reservoirs. Economically efficient carbon capture and storage (CCS) methods could become a powerful way of mitigating some of the climate damage that is being caused by the use of hydrocarbons in transport and power generation applications. A requirement of acceptable CCS programmes is that it be possible to monitor CO_2 disposal at these sites to provide assurances that the gas remains captured in the storage reservoirs.

The first two examples presented in the text that follows show projects where repeat gravity surveys have been used successfully. The last section describes a feasibility modelling study to assess the feasibility of using gravity and gravity gradients to monitor steam floods in heavy oil reservoirs.

Pau Reservoir Monitoring

Shell, Total and Statoil (now Equinor) funded a research project to test the use of time-lapse micro-gravity to monitor a gas storage buffer facility near Pau in south-western France (Bate, 2005). The project was operated by ARK Geophysics between January 2003 and June 2004.

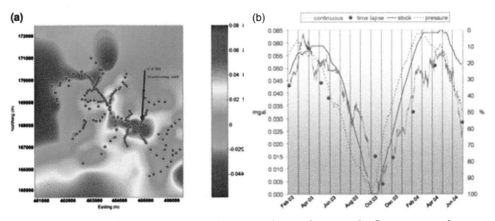

Figure 5.32 (a) Maximum observed changes in gravity over the Izaute reservoir from October 3 to April 4, which was a period of gas extraction. Amplitude on the colour bar is in mGal. (b) Diagram showing the time-lapse gravity (blue dots), continuous gravity (blue trace), gas stock levels (red trace) and reservoir pressure (red dashed trace) at a monitoring well over a 16-month period. As the reservoir is depleted, the gas/water contact rises, resulting in an increase in the time-varying gravity field. Note that the polarity of the gravity data has been inverted to facilitate easier comparison. [(a) After figure 2 and (b) figure 3 in Bate, 2005]

The Izaute gas storage reservoir is used as a buffer to manage local gas supplies. The storage is depleted in the winter months and filled through the summer. The objective of this trial was to investigate whether time changes in the observed gravity field could practically be measured and if so, whether the observations correlated to known changes in gas stock levels.

Gravity data were recorded continuously at two locations (one over the reservoir and the other sufficiently far away to serve as a control) and other locations were recorded approximately monthly over an 18-month period. The results are shown in Figure 5.32a and b (Bate, 2005). Figure 5.32a shows the maximum change in gravity for a whole cycle of charge/discharge and is in good agreement with expectations from modelling. Figure 5.32b shows a persuasive correlation of changes in gravity with level of gas storage and changes in reservoir pressure through time. Quite clearly the local gravity anomaly is responding to the production/storage cycle of gas as expected from models. This is an elegant example of use of a classical (and cost-effective) technology to accurately monitor subsurface changes that create density variations with time.

CO₂ Injection at the Sleipner Field, Offshore Norway

The Sleipner gas condensate field was put on production in 1996 (Furre et al., 2017). The gas in Sleipner has a high CO_2 content (approximately 9%) and the

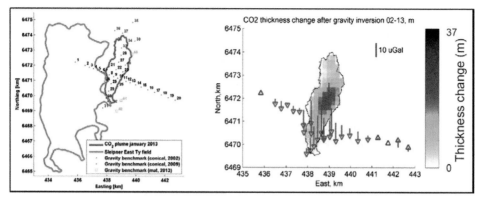

Figure 5.33 Left: The gravity survey layout, with outlines of the CO2 plume and the deeper Sleipner East Ty gas-condensate reservoir. Right: Inverted CO_2 thickness change based on the gravity change from 2002 to 2013, along with the CO_2 outline from seismic in 2013. Red arrows show decrease in measured gravity, blue arrows increase. The signal from Ty Fm. has been removed along with other signals mentioned in the text. (After figure 3 from Furre et al., 2017)

operator was required o strip this from the produced hydrocarbons and dispose of it safely. Sleipner became the first commercial CCS project.

The operator decided to re-inject the CO_2 into a sandstone unit of the Early Pliocene age Utsira formation. Approximately 0.9 million metric tons of CO_2 have been injected since 1996. Since the Norwegian state regulator required assurances that the CO_2 remained sequestered in this formation, the operator undertook a programme of regular remote monitoring using primarily repeat seismic and gravity data acquisition.

Furre et al. (2017) show a number of time-lapse seismic images that demonstrate the reservoir distribution changing with time but remaining within the confines of the closure at the Utsira level. As well as 4D seismic (10 repeat surveys) the programme also included regular recordings of gravity measurements using permanent gravimeters placed on the seabed (four surveys).

The first gravity survey was conducted in 2002, with repeat surveys in 2005, 2009 and 2013. The gravity anomaly resulting from CO_2 injection from the 2013 repeat survey is shown in Figure 5.33. The close match with the seismic anomaly is good demonstration that the local gravity field change is due to the CO_2 storage.

Use of Gravity Gradients in Reservoir Monitoring and CCS

The above two examples illustrate the feasibility of detecting reservoir fluid changes using gravity (g_z) measurements. An interesting question is whether *gravity gradients* can add any significant value. Would there be any added benefits to be gained from measuring or calculating changes in a gravity gradient tensor field caused by fluid flow?

The scale of the g_z changes created by substitution of brine/methane/oil/CO_2 is approximately a few to 100 μGals. The two examples described earlier show that it is possible to detect such changes using *stationary* gravimeters. Unfortunately, dynamic systems such as the Lockheed Martin FTG mounted on moving platforms are not capable of detecting such small signals: the signal-to-noise levels inherent in deploying accelerometers on a moving platform are too low for detecting such small signals. Hence today any application of gravity gradients would need to rely on using fixed gravimeters and then deriving the gravity gradients (G_{zz}, G_{zx}, etc.) mathematically.

An example of applying gravity gradients to the monitoring problem is given by Elliott and Braun (2016), who present a modelling case study of a steam flood into a heavy oil/bitumen reservoir in the Alberta Basin. Such heavy oil reservoirs are often produced with the help of steam injection.

In this example the reservoir is at an approximately 200 m depth and expected density changes are in the range of 0–0.3 g/cm^3 when displacing oil with steam. In this study the authors use a model with a single pair of wells (injector and producer) and show that both gravity (g_z) and gravity gradients (G_{zz}, G_{zx}, etc.) are able to detect a density change of the expected scale (see Figures 5.34 and 5.35).

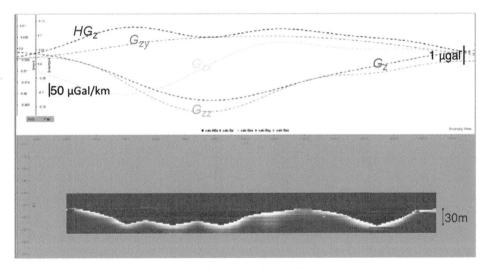

Figure 5.34 Cross section of the density variations in the reservoir and the gravity (G_z) and gravity gradient responses (G_{zz}, G_{zx}, G_{zy}, HG_z), after 5 years of production (the cross line section is taken from the middle of the reservoir). (After figure 7 from Elliott and Braun, 2016. Courtesy of CSEG Recorder)

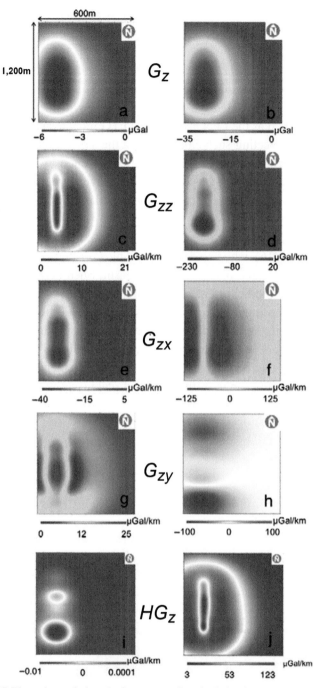

Figure 5.35 Plan view of signals for one well pair (injection and production). (a) G_z at 1.25 years. (b) G_z at 5 years. (c) G_{zz} at 1.25 years. (d) G_{zz} at 5 years. (e) G_{zx} at 1.25 years. (f) G_{zx} at 5 years. (g) G_{zy} at 1.25 years. (h) G_{zy} at 5 years. (i) HG_z at 1.25 years. (j) HG_z at 5 years. (After Elliott and Barun, 2016, figure 8. Courtesy of CSEG Recorder)

Here the gradients are calculated mathematically from the measured scalar gravity. This is a common approach in gravity surveying: differentiating the field presents a higher resolution image.

The impact of displaying gravity gradients is more clearly seen in Figure 5.35, which shows that gradient displays show a much sharper image of density changes due to the steam flood between the two wells.

This example shows a proof of concept (at a feasibility level). In principle, gravity gradients could provide a valuable measure of density changes resulting from production and/or injection (in oil and gas field operations or in sequestration assurance or CCS).

This study was based on deploying a superconducting gravimeter (iGRAV; see Elliot and Barnes, 2016) which is static while measurements are being taken. Such instruments are expensive and difficult to manage in the field, which creates serious problems in the context of commercial deployment over many fields.

Absent a device that can be deployed on a mobile platform (such as the FTG) what is needed is a cheap, robust gravimeter than can be deployed in large numbers over a field area. A possible candidate for such an instrument is a MEMS (microelectromechanical systems) gravimeter (see Middlemiss et al., 2016). MEMS accelerometers are routinely installed in modern smartphones and are thus fairly cheap devices. Middlemiss et al. (2016) describe the effective deployment of such devices in a small, robust gravimeter that can be used in many applications. One such application is in the Newton-g research project to monitor the behaviour of the Mount Etna volcano (www.newton-g .eu/). The plan here is to deploy approximately 40 MEMS gravimeters on Mount Etna for an extended period of time and monitor the gravity changes created by migration of magma within the volcano as part of an early warning system.

The prototype MEMS gravimeters are considerably cheaper than conventional devices. It is conceivable that following successful applications in the field, the cost of these small devices could reduce significantly (perhaps by a factor of 10 or more?) through mass production. In this eventuality it would be possible to deploy (possibly permanently) a large number of MEMS devices over a reservoir for continual observation of gravity changes. One could then use either gravity (g_z) or a derived gravity gradient tensor field (G_{zz}, etc.) to monitor accurately changes in reservoir fluid saturations. A great advantage from using cheap fixed systems is that this improves repeatability considerably when taking differences.

With instruments such as the MEMS it becomes highly feasibly both technically and economically to use gravity and gravity gradients as a tool for reservoir monitoring (including CCS). Something to follow as testing of this and similar devices develops over the next few years.

5.5.3 Summary

A comprehensive comparison of the relative merits of gravity and gradiometry is given in Table 4.2 of Chapter 4. The main additional benefits of AGG are

- Improved accuracy of locating structural fabric at prospect scale
- Reducing the problem of structural aliasing inherent in 2D seismic data interpretation

It is these two exploration benefits that have driven the growth of the application of FTG/AGG technology. As with conventional gravity, gradiometry information is also obviously useful for building better seismic velocity models. This enables improvements in seismic imaging – possibly at higher resolutions depending on depth.

FTG/AGG is best used in combination with other tools and knowledge. It is very seldom that any single geophysical tool is sufficient to be used in a standalone mode for the reduction of uncertainty and risk.

In common with other geophysical tools, FTG/AGG surveys need very careful design – this is especially so given that it is a relatively young technology and hence there is limited general experience of the controls on data quality. Processing of gradiometry data is also a developing science and is very much in the realm of the specialist. For example, there are differences in the data acquired using the two variants of gradiometry (the FTG and Falcon AGG systems) and this has implications for both survey design and processing approaches.

The choice of whether to use gradiometry or conventional gravity is very much project specific and depends largely on the depth of the target. The ideal would be to acquire both conventional and gradiometry data in order to *measure* both long and short wavelengths. This is not usually practical on cost grounds; however, the innovative eFTG solution is a highly cost-effective alternative to running two separate surveys.

What is clear is that the advent of instruments that can economically acquire a broad band measure of the gravity field, with a 'sideways looking' capability has introduced a significant new tool to the explorer's kit. Intelligence and care need to be applied for effective deployment and this is usually done through modelling and feasibility studies (fortunately these are quick and cheap to carry out).

It is clear that this technology has potential across the whole oil and gas exploration and production value chain (from prospect evaluation through to reservoir monitoring). The improvements in resolution and signal/noise quality may even provide a basis for application to reservoir monitoring and CCS assurance.

Acknowledgements

I wish to thank Paul Versnel, Dr. Gary Barnes and Dr. Jonathan Watson for many helpful comments and their feedback on this chapter. I also wish to thank them (and other colleagues who worked for ARK Geophysics and ARKex Geophysics) for introducing me to the subject of gravity gradiometry and for the many insights into the application and use of this remarkable technology.

References

Barnes, G., 2012. Interpolating the gravity field using full tensor gradient measurements. *First Break*, **30**, 97–101.

Barnes, G., 2018. The gravity module assembly used in airborne full tensor gradiometry surveys. *Geophysical Prospecting*, **67**. DOI: 10.1111/1365-2478.12707.

Barnes, G. and Lumley J., 2010. Noise analysis and reduction in full tensor gravity gradiometry data. In R. J. L. Lane, ed., Airborne Gravity 2010. Abstracts from the ASEG-PESA Airborne Gravity 2010 Workshop. Published jointly by Geoscience Australia and the Geological Survey of New South Wales, Geoscience Australia Record 2010/23 and GSNSW File GS2010/0457, 21–27.

Barnes, G. and Lumley, J., 2011. Processing gravity gradient data. *Geophysics*, **76**(2), March–April, I33–47. DOI: 10.1190/1.3548548.

Bate, D., 2005. 4D reservoir volumetrics: A case study over the Izaute gas storage facility. *First Break*, **23**, 69–71.

Colombo, D. and McNeice, G. W., 2013. Subsalt imaging with full-tensor electromagnetics in the Red Sea. In *Extended Abstract, EAGE Conference 2013*, London, DOI: 10 .3997/2214-4609.20130733.

Davies, M. A. and Martin, J., 2010. Application of gravity gradiometry in salt basin modelling. KAZGeo 2010: *Proceedings of 1st EAGE International Geosciences Conference for Kazakhstan*. EAGE EarthDocs.

DiFrancesco, D., 2007. *EGM 2007 International Workshop, Innovation in EM, Grav and Mag Methods: Anew Perspective for Exploration*, Capri, Italy, 15–18 April 2007.

DiFrancesco, D., 2011. The growth of airborne gravity gradiometry – and challenges for the future. In*Twelfth International Congress of the Brazilian Geophysical Society*

DiFrancesco, D., Meyer, T., Christensen, A. and Fitzgerald, D., 2009. Gravity gradiometry – today and tomorrow. In *11th SAGA Biennial Technical Meeting and Exhibition Proceedings*, Swaziland, 16–18 September 2009, 80–3. DOI: 10.3997/2214-4609-pdb.241.difrancesco_paper1.

DiFrancesco, D. and Talwani, M., 2002. Time-lapse gravity gradiometry for reservoir monitoring. In *Expanded Abstracts, 72nd Annual International Meeting, SEG*, 787–90.

Dransfield, M., 2010. Conforming Falcon® gravity and the global gravity anomaly. *Geophysical Prospecting*, **58**(3), March, 469–83. DOI: 10.1111/j.1365-2478.2009.00830.x.

Dransfield, M. H. and Christensen, A. N., 2013. Performance of airborne gravity gradiometers, Fugro Airborne Surveys. *The Leading Edge,* 909–22.

Elliott, J. E. and Braun, A., 2016. Gravity monitoring of 4D fluid migration in SAGD reservoirs: Forward modelling. *CSEG Recorder*, 41(01), 16–21.

Furre, A.-K., Eiken, O., Alnes, H., Vevatne, J. N. and Kiaer, A. F., 2017. 20 years of monitoring CO_2-injection at Sleipner. *Energy Procedia*, **114** (2017), 3916–26. DOI: 10.1016/j.egypro.2017.03.1523.

Houghton, P., Nuttall, P., Cvetkovic, M. and Mazur, S., 2014. The role of potential fields as an early dataset to improve exploration in frontier areas. *First Break*, **32**(4), April. DOI: 0.3997/1365-2397.32.4.74382,

Jackson, D., Helwig, J. H., Dinkelman, M. G., Silva, M. and Protacio, J. A. P., 2013. Integration of 2D seismic, gravity gradiometry, and magnetic data on a passive Margin – NE Greenland. In *EAGE Conference 2013*, Tu 09 04. DOI: 10.3997/2214-4609.20130443.

Jorgensen, G. J., Kisabeth, J. L. and Routh, P., 2006. The role of potential fields data and joint inverse modeling in the exploration of the Deep Water Gulf of Mexico Mini-Basin Province. Bell Geospace. www.bellgeospace.com/doc/frontiers_ed5_main_Jorgensen_lr.pdf .

Middlemiss, R. P., Samarelli, A., Paul, D. J., Hough, J., Rowan S. and Hammond, G. D., 2016. Measurement of the Earth tides with a MEMS gravimeter. *Nature*, **53131**, March. DOI:10.1038/nature17397.

Moore, D., Chowdhury, P. R. and Rudge, T., 2012. FALCON airborne gravity gradiometry provides a smarter exploration tool for unconventional and conventional hydrocarbons: case study from the Fitzroy Trough, onshore Canning Basin. In T. Mares (ed.), *Eastern Australasian Basins Symposium IV: Petroleum Exploration Society of Australia*, Special Publication, CD-ROM. www.spgindia.org/10_biennial_form/P114.pdf

Murphy, C. A., 2004. The Air-FTG™ airborne gravity gradiometer system, in R.J.L. Lane, ed., Airborne Gravity 2004. *Abstracts from the ASEG-PESA Airborne Gravity 2004 Workshop. Geoscience Australia Record*, 2004/18, 1–5.

Murphy, C. A., Mumaw, G. R. and Zuidweg, K., 2005. Regional target prospecting in the Faroe-Shetland Basin area using 3D-FTG Gravity data. In *EAGE 67th Conference & Exhibition*, Madrid, Spain, 13–16 June 2005. DOI: 10.3997/2214-4609-pdb.1.P503.

Nabighian, M. N., Ander, M. E., Grauch, V. J. S., et al. 2005. Historical development of the gravity method in exploration. *Geophysics*, 70(6), November–December; 63 ND–89ND, DOI: 10.1190/1.2133785.

O'Brien, J., Rodriguez, A., Sixta, D., Davies, M. A. and Houghton, P., 2005. Resolving the K-2 salt structure in the Gulf of Mexico. *The Leading Edge*, 24(4), 404–9. DOI: 10.1190/1.1901394.

Pengyu, L. and Guoqing, M., 2015. Balanced gradient methods for the interpretation of gravity tensor gradient data, *Journal of Applied Geophysics*, **121**, October, 84–92. DOI: 10.1016/j.jappgeo.2015.07.011.

Price, A. D., Cacheux, A., Chowdhury, P. R., Shields, G., Weber, J. and Yalamanchili, R., 2013. Airborne gravity gradient acquisition for oil exploration in Uganda. In *Extended Abstract EAGE London Conference* 2013. DOI: 10.3997/2214-4609.20130123.

Protacio, J. A., Watson, J., Van Kleef, F. and Jackson, D., 2010. The value of integration of gravity gradiometry with seismic and well data: An example from a frontal thrust zone of the UAE-Oman Fold Belt. Presented at EAGE Barcelona 2010. DOI: 10.3997/2214-4609.20149920.

Saad, A. H., 2006. Understanding gravity gradients: A tutorial. *The Leading Edge*, **25**, 942–9. DOI: 10.1190/1.2335167.

Sarkawi, I., Abeger, G., Tornero, J. L. and Abushaala, E., 2007. The Murzuq Basin, Libya - A Proven Petroleum System. In *Extended Abstracts, EAGE 3rd North African/ Mediterranean Petroleum & Geosciences Conference and Exhibition* Tripoli, Libya, 26–28 February 2007. DOI: 10.3997/2214-4609.20146465.

6

Marine Electromagnetic Methods

LUCY M. MACGREGOR

6.1 Introduction

Using measurements of resistivity derived from electrical or electromagnetic (EM) methods to understand the properties of the subsurface is not new. Indeed, some of the earliest geophysical methods on land measured resistivity, and this was followed shortly afterwards by offshore measurements using DC resistivity arrays (e.g., Schlumberger et al., 1934; Francis, 1977). Today, resistivity is a key measurement in well logging. The reason for this is illustrated in Figure 6.1, which shows a well log suite penetrating a clean reservoir sand saturated with oil.

Well log interpretation is by its nature an integrated process. A petrophysicist analysing a well log suite takes physical measurements of the earth and uses them to interpret the lithology and fluid properties at the well bore. The right five tracks in Figure 6.1 illustrate the effect of fluid saturation on seismically measurable properties. In each case the black curve shows the measured in situ response when the reservoir sand is oil saturated whereas the blue curves show the equivalent response for the water saturated case. The difference is relatively small. Contrast this with the resistivity response to the oil saturated sand shown in the fourth track (red curve). The presence of hydrocarbon in the sand increases the measured resistivity by two orders of magnitude. This large sensitivity to fluid saturation makes resistivity a vital component of any well log interpretation, providing complementary information to that obtained from the other logs.

Information on subsurface lithology and fluid properties is often needed away from the well bore. For this, remote sensing geophysical methods are needed. Seismic is the most commonly applied geophysical method and is an extremely powerful way of probing the earth: It is generally applicable, can provide high-resolution images of structure and stratigraphy under many circumstances and with careful analysis and calibration can provide information on subsurface properties. However, like all geophysical methods it has its limitations. In some situations,

169

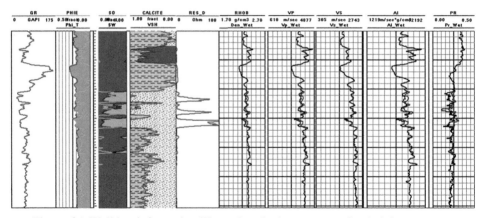

Figure 6.1 Well log information illustrating the importance of resistivity measurements. Tracks from left to right are gamma ray, porosity, fluid saturation, lithology, resistivity, density, P-wave velocity, S-wave velocity, acoustic impedance and Poisson's ratio. In the right five tracks the black curve shows the in situ oil saturated case, and the blue curve shows the equivalent response if the reservoir sand were water saturated. (From MacGregor and Cooper, 2010)

seismic imaging of structure may be complicated. For example, high-velocity or highly heterogeneous layers such as salt or basalt may obscure deeper structure. Interpretation of subsurface properties may also be ambiguous based only on seismic data. For example, it is often hard (and sometimes impossible) to determine fluid saturation reliably using only seismic measurements. This 'fizz gas' problem, where low-saturation gas cannot be distinguished from commercial pay, is well known and has led to many disappointing well results. In both of these cases, incorporating resistivity measurements into the interpretation can improve the result. EM methods provide a way of measuring subsurface resistivity and thus complement more conventional seismic approaches.

6.2 Overview of Marine Electromagnetic Approaches

Marine EM methods have been in use for a long time. Reviews of the technology are given by, for example, Chave et al. (1991), Edwards (2005), Constable and Srnka (2007), Constable (2010), and Key (2011), and only a brief overview will be given here. Development of marine Controlled Source Electromagnetic (CSEM) methods began in the late 1970s as a tool to map deep crustal resistivity structure at a large scale (Young and Cox, 1981; Constable and Cox, 1996). The method evolved throughout the 1980s and 90s through a number of surveys investigating the crustal scale structure of marine volcanic and hydrothermal systems (Evans et al., 1994; MacGregor et al., 1998, 2001). Even in these early surveys the importance of close integration with seismic methods was understood. Seismic information was used to provide structural constraint and CSEM derived resistivity information integrated to constrain fluid volumes and properties (Sinha et al., 1998).

Over the same period the method was also applied to the characterisation of marine gas hydrates (e.g., Edwards, 1997; Yuan and Edwards, 2000; Schwalenberg et al., 2005). Although seismic can in some circumstances identify the base of the hydrate stability zone, characterised by a bottom simulating reflector (BSR) from which the presence of hydrates can be inferred, it is hard on the basis of seismic alone to quantify the hydrate saturation. Resistivity measures can provide complementary information in this situation to assist the characterisation process.

As early as the mid-1980s applications to conventional hydrocarbon exploration (as opposed to hydrate exploration) were considered by oil majors, and a patent for the use of marine CSEM for detection of hydrocarbon reservoirs was granted to Exxon in 1986 (Srnka, 1986). Although the possibility of using CSEM in exploration looked promising, economic conditions prevented its application in practice.

In a parallel, but related development, during the 1990s marine magnetotelluric (MT) was under development for structural mapping (Hoversten et al., 1998; Constable et al., 1998), and a number of industry funded surveys targeting salt mapping were undertaken (e.g., Hoversten et al., 2000; Key et al., 2006). The receiver technology developed for the MT application is equally applicable to CSEM studies (Constable, 2013) and forms the basis of the node based receiver technology used today. Similarly CSEM was being investigated as an aid to subbasalt imaging (MacGregor and Sinha, 2000; MacGregor, 2003), under the industry-funded LITHOS consortium at the University of Cambridge.

By the end of the 1990s the equipment for marine CSEM and MT acquisition developed at the Scripps Institution of Oceanography in the United States and the University of Cambridge in the United Kingdom (this group moved to the National Oceanography Centre in Southampton in early 2000) was mature and reliable. Processing and interpretation algorithms had also been developed allowing data to be inverted in 1D and 2D to provide resistivity sections that were interpretable alongside seismic and other information. In addition, the search for hydrocarbons had moved to deeper water, where CSEM methods had traditionally been applied, and where well costs were significant, making successful de-risking of prospects a priority. This provided the critical economic driver that led to the first trial of the marine CSEM method for reservoir mapping in 2000. The collaborative research survey was funded by Statoil (now Equinor) and used equipment and expertise from Scripps in California and the National Oceanography Centre, UK. The target was a field in Angola, and results showed clearly that for this field at least, CSEM could indeed detect the high resistivity associated with a hydrocarbon accumulation (Ellingsrud et al., 2002). This initial survey was followed closely by further surveys funded by ExxonMobil.

The period following the first proof of concept survey saw rapid expansion in the marine EM industry. Several contractors were formed and large volumes of data collected. Marine CSEM in particular was marketed as the 'new' 3D seismic that could more reliably find hydrocarbons. Reality intruded in 2008 when there was a dramatic reduction in market size. A number of reasons for this can be postulated: First, over-estimation of future market size led to over-capacity in the CSEM acquisition market. Second, a combination of exaggerated marketing claims coupled with disappointing results led the industry to become sceptical of the technology and the value it could deliver. Finally, the importance of careful integration of CSEM or MT derived resistivity with seismic and well log information, and the importance of electrical anisotropy in interpreting results were not well understood.

Nevertheless, marine EM remains a valuable tool. If data are acquired and interpreted carefully, the resulting resistivity remains a valuable measurement that provides complementary information to that obtained from more traditional geophysical approaches.

6.3 Resistivity in the Earth

An important point to remember is that EM methods measure resistivity, and not hydrocarbons (or anything else) directly. The resistivity of earth materials varies over many orders of magnitude (Figure 6.2). Importantly, replacing conductive pore fluids, with resistive hydrocarbon can result in a large increase in resistivity (often by an order of magnitude or more), and it is this change that is the target for many marine EM surveys over potentially hydrocarbon bearing prospects.

However, a measurement of high resistivity does not guarantee the presence of hydrocarbons, commercial or otherwise. Many geological features are resistive. For example, tight sands, carbonates, volcanics or salt stringers may all have a resistivity signature that is identical to that of a hydrocarbon saturated sand. As with any other geophysical measurement, resistivity must be interpreted carefully, alongside seismic or other information, within a calibrated rock physics framework in order to determine the underlying rock and fluid properties.

Electrical anisotropy is a key consideration in marine EM surveys. Electrical anisotropy is common in sedimentary sequences with ratios of vertical:horizontal resistivity of 2:1 widespread, and values up to 10:1 observed in places (Klein et al.,

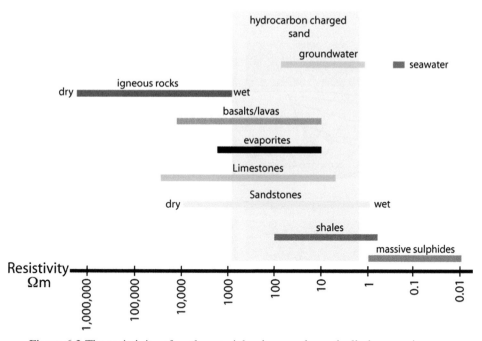

Figure 6.2 The resistivity of earth materials, shown schematically here, varies over many orders of magnitude, and depends on lithology, fluid content and fluid distribution, among other factors.

1997; Ellis et al., 2010; MacGregor et al., 2012; Colombo et al., 2013; Gabrielsen et al., 2013). The anisotropy is caused by a combination of macro- and micro-effects including layering, fracturing and grain scale alignments. In a regional study in the Barents sea, the anisotropy ratio was found to be variable with the primary control being uplift: The greater the paleo-burial depth, the greater the anisotropy, with grain alignment likely to be the controlling mechanism (Bouchrara et al., 2015; Ellis et al., 2017).

Such significant levels of anisotropy constitute a first-order effect in marine EM analysis and interpretation. It is therefore important to correctly account for anisotropy in survey planning (Ellis et al., 2011) to avoid suboptimal results. In data inversion and interpretation, failing to account correctly for electrical anisotropy can lead to artefacts in the result (Ramananjaona and MacGregor, 2010; Ramananjaona et al., 2011).

The majority of approaches assume that the earth exhibits vertical transverse isotropy (VTI) in which the horizontal resistivity is the same in all directions, but different from the vertical resistivity. This type of anisotropy would be exhibited by, for example, a stack of finely stratified horizontal layers. Recently there have been studies into the effect of tilted transverse isotropy (TTI) in which the

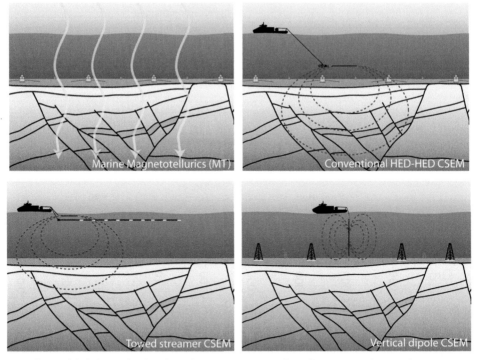

Figure 6.3 Schematic showing the commonly used marine EM methods. The most common source in marine CSEM is the horizontal electric dipole (HED), although vertical electric dipoles (VED) are also used. (From MacGregor and Tomlinson, 2014)

resistivity is constant in a plane, with a different resistivity normal to that plane, where this normal is tilted with respect to the vertical (Davydycheva et al., 2013). In areas of steeply dipping stratigraphy, for example, in thrust belts or on the flanks of salt diapirs, including TTI in inversion and interpretation workflows improves the robustness of the result (Hansen et al., 2016, 2018).

6.4 Practical Applications

Marine EM methods fall into two categories. Natural source methods use EM fields generated in the earth's atmosphere and ionosphere as a source. In controlled source methods, the source fields are generated artificially by a source that is under the control of the survey geophysicist (Figure 6.3). Constable (2013) provides a review of source and receiver acquisition technology used in these methods.

Table 6.1 provides a comparison of controlled source and natural source methods, which are discussed in more detail in the text that follows.

Table 6.1 *Comparison of natural and controlled source EM methods*

Natural source methods: MT	Controlled source methods: CSEM
Naturally generated plane wave source	Artificially generated dipole source
Resolution from studying response as a function of frequency and lateral position	Resolution from studying received fields as a function of frequency, source receiver separation and geometry
Relies predominantly on horizontal current flow: very insensitive to thin resistive layers	Source induces both horizontal and vertical current flow: sensitive to thin resistors
Frequency range typically 0.00001 Hz–1 Hz	Frequency range typically 0.01 Hz–10 Hz
Can be used to determine resistivity structure to tens of kilometres below seafloor	Typically sensitive to depths of c. 3–4 km below mudline (dependent on structure)
Most applicable to mapping large-scale resistivity structure; salt mapping; basalt mapping (when basalt is thick); assessment of depth to basement	Most applicable to mapping resistive structure; basalt mapping (when basalt is thin); characterisation of hydrocarbon accumulations

In both cases resolution depends on depth of burial, and, in general, lateral resolution will be better than vertical resolution.

6.4.1 Natural Source Methods: Marine Magnetotellurics

The marine MT method uses naturally generated EM fields as a source (Figure 6.3, top left). The frequency-dependent transfer function calculated between the magnetic and orthogonal electric fields gives an earth EM impedance at a range of frequencies, typically between 0.0001 Hz and 10 Hz for marine applications. The impedance magnitude, usually expressed as an apparent resistivity, and phase as a function of frequency can be used to determine electrical resistivity as a function of depth in the subsurface. By measuring this impedance across an array of receivers, lateral changes in resistivity can also be determined.

The MT method has been used for many years as an aid to petroleum exploration on land (see, e.g., Vozoff, 1972, or Orange, 1989). However, the method was thought to be less applicable offshore, since the conductive ocean layer acts as a screen, attenuating all but the lowest frequency signals, which are not sensitive to structure at depths relevant to petroleum exploration. This thinking changed in the mid-1990s with the development at the Scripps Institution of Oceanography of broadband marine MT receivers (Constable et al., 1998) capable of measuring signals in the 0.001 Hz–1 Hz range in water depths of a kilometre or more. It is worth noting that these receivers, developed primarily for marine MT, can also be used to record CSEM signals (discussed later in the chapter), and were used extensively in early applications of CSEM to hydrocarbon exploration.

Marine MT is particularly useful for large-scale basin reconnaissance in areas where seismic methods perform poorly (examples are given in Sections 6.8.1 and 6.8.2). Initial surveys focused on mapping salt structures in the Gulf of Mexico (Hoversten et al., 2000; Key et al., 2006) and results demonstrated that salt structures, and in particular base salt, could be mapped using MT in situations in which seismic imaging was poor or absent. Since these initial surveys, marine MT has been applied extensively, including further surveys to map salt in the Gulf of Mexico (e.g., de Stephano et al., 2011), Barents Sea (Hokstad et al., 2011; Stadtler et al., 2014) and Red Sea (Colombo et al., 2013).

Similarly, marine MT has found application in areas where potentially prospective sediments are obscured by high-velocity or seismically heterogeneous layers such as basalt. The method has been applied offshore India and West of Shetland to map basalt structure (e.g., Jegen et al., 2009; Panzer et al., 2016). In these cases resistivity information can be used to construct more accurate velocity models, leading to improved seismic imaging (de Stephano et al., 2011; Panzer et al., 2016). Applications to mapping sediment thickness and depth to basement in frontier areas have also proved valuable (Karpiah et al., 2019).

Acquiring MT data is logistically simpler than acquiring controlled source EM data. The only requirement is that receivers capable of measuring orthogonal components of the electric and magnetic field are deployed on the seafloor and left to record the natural signals. The length of 'listening time' required varies with water depth and target structure, however deployment times in the range of 24–48 hours are typical.

However, the MT method has limitations. The natural source fields used as a source are just that: natural. Their strength fluctuates depending on solar and atmospheric activity, potentially affecting data quality. The geometry of the MT source field also affects the type of structure to which the method is sensitive. The source fields are predominantly horizontal, and therefore induce largely horizontal current flow in the subsurface. As a result, although the resulting impedances are extremely sensitive to large bodies such as salt diapirs or thick basalt, the method is notoriously insensitive to thin resistive structure, for example, thin basalt layers, horizontal salt stringers, or more importantly hydrocarbon-bearing reservoirs. To resolve such structures, controlled source EM methods must be used.

6.4.2 Controlled Source Methods: Marine Controlled Source Electromagnetic Surveying

CSEM methods use artificially generated source fields to induce fields in the subsurface. The resulting response is detected and recorded by electric and/or magnetic field receivers deployed on the seafloor or towed behind the source (Figure 6.3).

The requirement to mobilise, deploy and tow a source makes CSEM data in some respects more complex logistically to acquire than MT. However, for many applications the benefits are clear. First, because the source is under the control of the geophysicist, survey parameters such as transmission frequency and source–receiver offset and geometry can be tailored to ensure optimum sensitivity to the targets of interest. Second, and more important for petroleum applications, the geometry of the fields of an electric dipole is such that both horizontal and vertical current flow are induced in the sub-surface. It is this vertical component, absent in MT methods, that makes the marine CSEM method sensitive to thin resistive structures that can be indicative of hydrocarbon accumulations.

Industry applications are dominated by the use of electric dipole–dipole systems, in which both the source and receiver is an electric dipole. There are three main approaches currently available.

6.4.3 Conventional Nodal HED–HED Methods

CSEM using a horizontal electric dipole (HED) source towed behind a survey vessel, and autonomous deployed receivers measuring the electric field has been in use since the late 1970s (Cox, 1981; Young and Cox, 1981), and has changed little since then, with the main advances being in increased source power and lowered receiver noise floor (Constable, 2013; Nguyen et al., 2017). A schematic of this acquisition approach is shown in the top right panel of Figure 6.3.

The source is formed by a neutrally buoyant streamer, usually approximately 300 m long. The streamer supports two electrodes, and the source signal is formed by passing high current between these. The strength of the source is measured by its dipole moment: the product of the dipole length and transmitted current. Source currents of around 1500 A are employed in most systems, although source currents up to 7500 A have been reported, giving source dipole moments of up to 2,500,000 Am (Barker et al., 2012; Nguyen et al., 2017). The source transmits a coded tri-state waveform, which can be designed to optimise the frequency content of the resulting signal for sensitivity to the targets of interest (Mittet and Schaug-Pettersen, 2008; Myer et al., 2010). Frequencies in the range 0.01 Hz–50 Hz are typical. The source is towed 30–50 m above the seafloor to ensure good coupling of signals to the subsurface, while avoiding entanglement with or damage to seafloor infrastructure (Sinha et al., 1990). Source towing speeds are typically slow, around 1–2 knots, a result of the need to control a source vehicle close to the seafloor at the end of several kilometres of tow cable.

The receivers detect and record up to three components of the electric and three components of the magnetic field, although it is most usual to record just the horizontal electric and magnetic field components. Although it is primarily the electric field used in CSEM analysis, recording the magnetic field as well allows MT data to be acquired 'for free', during periods when the source is off or out of range of the receiver. These MT data can provide valuable complementary information on background resistivity structure. At the end of the survey the receivers are recovered to the survey vessel and the data downloaded for processing and interpretation. Data are analysed in the frequency domain and comprise the amplitude and phase of the signal at the specific frequencies transmitted. The (known) source signal is deconvolved during processing. By studying the variation in the measured amplitude and phase as a function of source–receiver separation, geometry and signal frequency, using a combination of forward modelling, inversion and hypothesis testing, the resistivity of the seafloor can be determined at scales of a few tens of metres to depths of about 3–5 km below mudline, depending on the structure and resistivity of the overburden.

6.4.4 Towed Streamer CSEM

More recently, a CSEM acquisition system in which both the source and receivers are towed at depths of 10–100 m below the sea surface has been developed (Mattsson et al., 2012; Engelmark et al., 2014). This is illustrated in the bottom left panel of Figure 6.3. The source in this case is an 800 m long horizontal electric dipole through which a 1,500 A current is transmitted, to give a dipole moment of 1,200,000 Am. The source signal is again a coded waveform comprising an optimised repeated sequence of 120 s duration, allowing energy to be distributed over a range of frequencies. Off periods between the transmissions of this sequence are used to characterise the noise levels in the system.

The receivers are built into a streamer that is also towed behind the survey vessel, giving maximum source receiver separations of 8 km in the currently available systems. The receiver array is towed at a depth of 100 m below sea surface. Receiver dipole lengths vary with offset, from 200 m at the near offsets, to 1,100 m at the longest offset to ensure signal-to-noise level is maintained as the signal attenuates with range. Signal to noise ratio is further enhanced by stacking the densely sampled receiver responses and using motion sensors distributed along the streamer to account for sudden changes in streamer velocity. Data are analysed in the frequency domain, with amplitude and phase as a function of source–receiver separation and frequency interpreted to determine resistivity to depth of about 2.5 km below mudline (again depending on structure).

The shallow tow depth of both the source and receivers allows acquisition speeds of 4–5 knots, leading to significant operational efficiencies over conventional CSEM acquisition. However, the shallow tow depth also limits operations of towed system to water depths less than about 500 m (McKay et al.,

2015). At greater depth the source signals are attenuated rapidly in the seawater layer, and this limits depth of penetration below the seafloor. A further limitation of towed streamer EM is that only a single data geometry is acquired: the inline geometry in which the source and receiver dipole axes are parallel and aligned along a line joining source and receiver. This geometry is optimally sensitive to structure that is resistive and thin compared to its depth of burial (MacGregor and Sinha, 2000; Ellingsrud et al., 2002). As a result, it is not a limitation per se; however, multi-azimuth data can in some circumstances be useful in constraining background structure and electrical anisotropy. However, the dense receiver sampling, and frequency coverage to some extent, mitigate this deficiency (Mattsson et al., 2013; MacGregor and Tomlinson, 2014), and extremely high-quality results have been achieved (e.g., Engelmark et al., 2014; Du et al., 2017).

6.4.5 Vertical Electric Dipole Methods

The final CSEM method in use in petroleum exploration today uses a vertical electric dipole (VED) for both the source and receivers (Holten et al., 2009; Helwig et al., 2019). This is illustrated schematically in the bottom right panel of Figure 6.3. The source comprises two electrodes, one placed on the seafloor and one positioned above it, 50 m below the sea surface. The total current transmitted is around 6,000 A (through a dual-source system). The dipole moment depends on the separation of the electrodes and hence on the water depth. Square wave pulses of alternating polarity are transmitted, separated by off-periods, during which the time domain response of the earth is measured at an array of seafloor receivers. The source remains stationary during transmission.

The receivers measure the vertical electric field using a vertical receiver dipole supported by a tripod. The receiver dipole is screened to avoid the effects of motional noise on the measured signals. Data are acquired and analysed in the time domain to determine the resistivity of the underlying seafloor.

The challenge with making this measurement is the small size of the vertical electric field in comparison to the horizontal electric field. As a result, only small tilt angles in the vertical source and receiver can be tolerated before the horizontal component dominates the response. However, a resistive earth structure can be detected at significantly shorter offset than in the horizontal electric dipole case (Cuevas and Alumbaugh, 2011), with the result that the vertical dipole system has the potential to improve lateral resolution of sub-seafloor resistivity structure compared to horizontal electric dipole systems. Good results have been achieved in a variety of settings (see, e.g., Alumbaugh et al., 2010; Helwig et al., 2013).

Table 6.2 gives a summary of the three commonly applied CSEM methods and the pros and cons of each. Note that the acquisition methods do not necessarily

Table 6.2 *Comparison of marine CSEM methods*

Method	Conventional HED/HED	Towed streamer CSEM	VED–VED CSEM
Source type	Horizontal electric dipole	Horizontal electric dipole	Vertical electric dipole
Source current/length	1,500 A–7,500 A/300 m	1,500 A/800 m	5,000 A/dipole length dependant on water depth.
Source coverage	Continuous	Continuous	Static transmission locations separated by c. 1–2 km
Receiver type	Autonomous seafloor deployed nodes. Two orthogonal components of horizontal electric and (optionally) magnetic field.	Towed streamer. Single inline component of electric field.	Autonomous seafloor nodes recording vertical electric field.
Analysis domain	Frequency domain	Frequency domain	Time domain
Receiver spacing	1–3 km	Continuous	1–3 km
Source line spacing.	Typically over the Rx lines and occasionally between lines (1–3 km)	1 km (prospect specific) – 20 km (regional studies)	Source stations typically between receiver locations.
Source-receiver offset/geometry	Offsets 200 m–15 km (typical), with multi-azimuth geometry	Offsets 200 m–8 km, single inline geometry.	Offsets 700 m–1.5 km azimuth less relevant for VED.
Tow speed	1–2 knots	4–5 knots	n/a: static source transmission locations.
Operational water depth range	c. 30 m–3.5 km Limited equipment rating and operations.	Water depths <500 m. Surface towed system only applicable in relatively shallow water.	Used in water depths up to c. 1 km to date. Limited only by complexity of operation as water depth increases.
Pros	Good areal and offset coverage. Multi-azimuth data acquired. Can operate in water depths of 20 m to 3.5 km.	Fast and efficient acquisition. Real time data QC. Dense receiver coverage across offsets.	Good lateral resolution of resistivity structure.
Cons	Little real time data QC. Time required to	Limited to 500 m water depth. Single	Relatively sparse sampling.

Table 6.2 (*cont.*)

Method	Conventional HED/ HED	Towed streamer CSEM	VED–VED CSEM
	deploy receiver array, and slow source tow speed can lengthen survey times.	field component data. Maximum offset currently 8km: may impact depth of penetration.	Operationally less efficient than horizontal electric dipole acquisition.
Applications	Exploration to development	Exploration to development.	Near field or step out exploration, appraisal to development.

Modified from MacGregor et al. (2019).
Examples later in this chapter illustrate the use of marine EM methods and the resolution that can be achieved.

compete under all circumstances. All have strengths and weaknesses, and the most appropriate should be chosen based on the geophysical question to be addressed, and the budget available.

6.5 Sensitivity and Survey Design Considerations

The sensitivity of both marine MT and CSEM methods to sub-seafloor structure must be considered when designing an acquisition campaign. Looking first at MT data, the left hand track of Figure 6.4 shows a simple 1D resistivity model, representing (for example), a thick anisotropic sedimentary layer overlying electrical basement. The water depth is assumed to be 1 km. Although the model is anisotropic, the MT response, at least in 1D, is sensitive only to the horizontal component of resistivity, the result of the predominantly horizontal current flow induced in the earth.

MT data are traditionally presented in the form of apparent resistivity and phase, calculated from the impedance tensor linking electric and magnetic fields. Apparent resistivity, proportional to the magnitude of the impedance tensor, is the resistivity of the uniform earth that would give the measured response on a frequency by frequency basis. Phase describes the lag between electric and magnetic fields (further details can be found in, e.g., Hoversten et al., 2000 and references therein). The MT apparent resistivity response of the 1D model is shown in black in the centre track of *Figure 6.4*. At short periods, which penetrate only the upper $1\Omega m$ layer, the apparent resistivity is close to 1 Ωm. At intermediate periods the apparent resistivity increases as the effect of the electrical basement starts to be felt. At the longest periods, the upper $1\Omega m$ layer is not resolved, and the response asymptotes towards 100 Ωm. Apparent resistivity as a function of

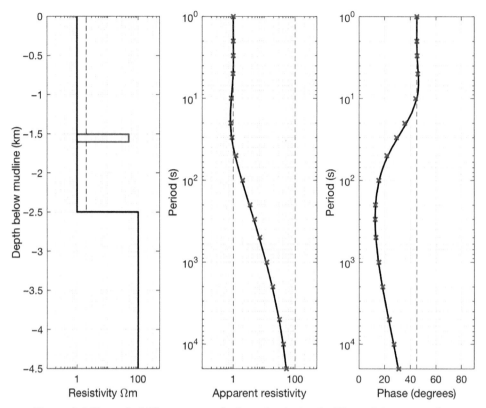

Figure 6.4 Example MT responses. Left track: A simple 1D model representing sediments over a more resistive basement (black curves). The overburden is anisotropic with a horizontal resistivity of 1 Ωm (solid line) and a vertical resistivity of 2 Ωm (dashed line). A reservoir of resistivity 50 Ωm and thickness 100 m is added at 1.5 km below mudline (red). The model is overlain by a water layer of thickness 1 km. Centre track: MT apparent resistivity response of the background model (black curve) and reservoir model (red). Right track: Phase response of the background model (black curve) and reservoir model (red).

period therefore provides a smoothed representation of resistivity as a function of depth. The phase response of a uniform structure is 45°, and therefore at short periods where only the upper layer affects the response, a phase of 45° is observed. The phase response deviates from this as the period increases and the deeper resistive structure starts to affect the response. At the longest ranges the phase asymptotes back to 45°.

Also shown in Figure 6.4 is the response of a model in which a thin resistive layer (100 m thick in this case) representing a hydrocarbon reservoir has been inserted into the model at a depth of 1.5 km. Neither the apparent resistivity nor the phase deviate significantly from the background response. MT data are extremely insensitive to thin resistive layers in the structure.

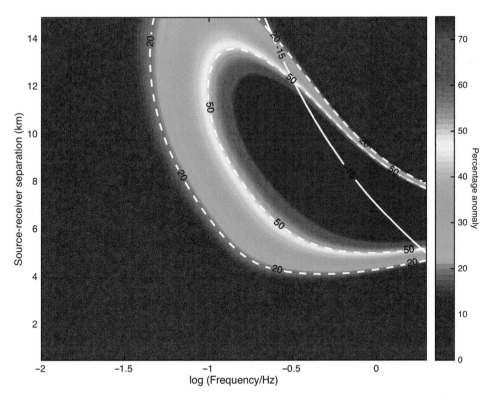

Figure 6.5 CSEM sensitivity analysis for the reservoir model shown in Figure 6.4. The percentage difference between the inline horizontal electric dipole response of the model containing the reservoir, and that of the background structure with no reservoir is contoured as a function of source receiver separation and source frequency. The dashed lines show the 20% and 50% contours. The solid white line shows the 10^{-15} V/Am^2 electric field strength contour and corresponds to a typically achievable noise floor. Signals above and to the right of this contour would not be detected above the noise.

This is not the case for marine CSEM data. Figure 6.5 shows the sensitivity of an horizontal electric dipole source with nodal receivers to the reservoir shown in the left hand track of Figure 6.4. The sensitivity is defined as the percentage difference in the measured inline electric field between the response of the model with the reservoir included, and the response of the background model (black lines in the left panel of Figure 6.4). CSEM data are sensitive to both horizontal and vertical resistivity, and correct characterisation of the anisotropy of the background structure is key to ensure correct interpretation of data and results (Newman et al., 2010; MacGregor and Tomlinson, 2014).

Sensitivity plots such as these provide a useful indication of the range of usable survey parameters. In the example shown in Figure 6.5, at very low frequencies (below about 0.01 Hz) the effect of the reservoir is small and observed only at long ranges. As a result, the resolution is likely to be poor. At high frequencies (above about 2 Hz in this case) the effect of the reservoir on the measure response is large, but signals are likely to be below the noise floor. The range of usable frequencies is therefore limited, in this case to about 0.1–1 Hz, and a waveform containing strong frequency components in this band would be optimum.

Figure 6.5 shows one example: in reality the sensitivity of the CSEM response to the subsurface structure is complex and dependent on a number of factors:

- Reservoir resistivity and thickness. The reservoir property that is constrained by a CSEM analysis is the transverse resistance of the reservoir: the vertically integrated resistivity (Constable et al., 2010). In the simple 1D model this equates to the resistivity-thickness product. The higher the transverse resistance, the larger the effect of the reservoir on the measured fields.
- Contrast between the reservoir and the surrounding sediments. Even if the transverse resistance is large, if there is very little contrast between the reservoir of interest and the surrounding sediments, sensitivity will be low.
- Resistivity of the overburden and underlying sediments. CSEM makes a measurement of a volume of the earth, and the measured response cannot be said to be caused by any particular feature. Therefore, the complete section must be considered when designing a survey. Optimum source frequency in particular depends strongly on the resistivity of the overburden (MacGregor and Tomlinson, 2014).
- Anisotropy in the over- and underlying sediments. Anisotropic effects can be large, and if not correctly accounted for can lead to erroneous interpretations.
- Presence in the section of other resistive features such as tight sands, volcanics, salts or carbonates.

The last three points in particular have a large effect not only on the choice of source transmission frequency but also on survey feasibility as a whole. Assuming a survey is feasible, an incorrect choice of survey frequency may render the resulting data insensitive to the properties of interest. It is therefore important when considering a CSEM survey to have an understanding of the background structure and anisotropy. Knowledge of the background structure is as important as knowledge of the potential reservoir properties: The background should be considered as much of a target as the reservoir interval of interest.

In areas devoid of well log information, construction of a background resistivity trend is difficult, and therefore uncertainties in assessing survey feasibility and establishing effective survey design parameters are large. Where standard well log derived resistivity information is available, this may be used to construct a suitable background model. However, the uncertainty in this background model remains large. The reason for this is electrical anisotropy. Standard induction log measurements in vertical well bores measure the horizontal component of the resistivity. The vertical component of resistivity, to which the inline electric field is primarily sensitive (Ramananjaona et al., 2011), is often many times higher. Correctly estimating, or predicting anisotropy in the background structure, is important when designing a survey. Background models must be constructed using best estimates of anisotropy derived from previous CSEM experience in a given area, three-component well logs where available, and experience in similar geological domains. Where significant uncertainty remains, contingency should be built into survey planning exercises to ensure that the resulting data will provide the required sensitivity across a range of background resistivity and anisotropy values.

For a successful survey it is important to ensure the data acquired are sensitive to the structure of interest. However, this is not sufficient: The ability of the analysis methods to be applied to recover the properties of interest must also be assessed. This 'recoverability' question was addressed by Key (2009) and MacGregor (2012a) for the case of a 1D structures, and by MacGregor and Tomlinson (2014) for 2D structures. As an example, if a potentially hydrocarbon bearing reservoir lies immediately above a salt diapir, it may be possible to find acquisition parameters for which the sensitivity of the data to the reservoir is high, assuming the background properties are known exactly. However, when interpreting survey data, it would be unlikely that the reservoir could be independently resolved from the larger resistive diapir beneath. Similarly, reservoir structures overlain by significant resistive features such as basalt, carbonate or salt, are unlikely to be resolved.

In all cases careful pre-survey modelling based on a geological and background information is required to establish, first, whether CSEM data is likely to add valuable information to the decision-making process, and second, determine the optimum acquisition strategy. However, some general considerations are illustrated in Figure 6.6, which shows a schematic summary of possible play types where CSEM may be considered:

- **Target A:** This represents a good target for CSEM surveying. It is relatively shallow in the section (perhaps less than 3 km below mudline) and located in relatively homogeneous background structure free from other resistors such as tight carbonates or volcanics. Many Tertiary reservoirs fall into this category, making them ideal for CSEM surveying, although there have been successes in many other geological settings with similar characteristics. Pre-survey modelling is required to optimise acquisition parameters before a survey is undertaken.

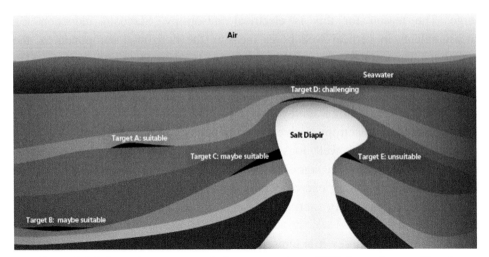

Figure 6.6 Schematic summary of play types where CSEM may be considered. Target A is likely to be suitable for CSEM. Targets B and C may be suitable depending on background structure, depth and reservoir properties. Target D is likely to be challenging because of the proximity of the reservoir to the salt diapir. Target E, which is overlain by salt, is unsuitable for current CSEM technologies. In all cases, however, careful pre-survey modelling should be undertaken before data are acquired (either new acquisition or purchase of multi-client data) to validate the suitability of the approach and optimise acquisition parameters.

- **Target B:** This may be a suitable target for CSEM surveying. It is more challenging than target A, because it is deeper in the section, and closer to resistive basement beneath. It is likely that overburden structure would be more heterogenous. Careful pre-survey modelling would be required to validate that this case was indeed suitable for CSEM methods, and to determine optimum acquisition parameters.
- **Target C:** This may also be a suitable target for CSEM surveying. The target reservoir is against the flank of a large resistor (a salt diapir) but displaced laterally from it. Once again, careful pre-survey modelling would be required to establish the accuracy with which such a prospect could be resolved and optimise the acquisition parameters for this purpose.
- **Target D:** This would be a challenging target, because the reservoir lies directly over the resistive salt diapir making it hard to separate the effect of the reservoir from the effect of the salt beneath. Depending on exact geometry, and exact proximity of the reservoir to the salt (or other resistor) it may be possible to provide some information on the target reservoir. Careful pre-survey modelling would be required to establish whether with the uncertainties likely to be present in the interpretation, this could still provide commercially useful information.

- **Target E:** This target would be unsuitable for CSEM surveying with currently available technology. Reservoir targets lying beneath significant resistors such as salt, basalt or thick resistive carbonate sequences would not be resolved using the CSEM method.

6.6 Basic Interpretation

The interpretation of marine EM data is a two-stage process, regardless of how the data were collected. In the first stage the data are processed and inverted to determine the resistivity distribution in the earth. In the second stage, the resulting resistivity must be combined with seismic, well log or other information to determine the underlying rock and fluid properties responsible for the observed resistivity variations. This process is illustrated with the examples in Sections 6.8.1 and 6.8.2.

Early interpretation of CSEM data relied on simple anomaly mapping approaches. Amplitude and phase data recovered from the receivers were normalised, using either data from a designated 'off target' receiver or the calculated response of a simple 1D model representing the background structure. Deviations between the data and the background response were interpreted to infer the presence of anomalous resistive structure in the subsurface (Ellingsrud et al., 2002; Hesthammer et al., 2012).

Such methods allow fast assessment of the data. Although sometimes useful for rapid reconnaissance and quality control (QC) of a data set, normalising approaches cannot be relied upon for detailed interpretation. There are several reasons for this. First, any anomaly identified is only anomalous relative to the chosen background model or data set. Since the background model is itself unknown, unambiguously determining the cause of any anomaly observed is challenging without further information. Second, although normalised anomalies highlight lateral variations in resistivity (relative to the chosen background) the depth of the body causing the variation cannot be easily determined. Finally, even if an anomaly is caused by a high-resistivity zone, this cannot be taken to be an unambiguous indicator of hydrocarbon: care must be exercised in establishing the underlying rock and fluid properties responsible for the observed response.

For both MT and CSEM data, a more robust approach to interpretation uses a staged forward modelling, inversion and hypothesis testing process to determine the resistivity distribution in the earth. 1D, 2D and 3D techniques all have their place. Interpretation approaches in general proceed in stages starting with simple 1D approaches, moving through 2D analysis and culminating in 3D analysis and inversion if required (Tseng et al., 2015).

1D forward modelling and inversion approaches assume that the earth is a 1D stack of layers. Each layer can be either anisotropic or isotropic. 1D forward models or inversions derived locally at receiver locations can be stitched together to derive resistivity pseudo sections or volumes. This process is similar to that often used in seismic inversion where locally 1D models are used in the inversion and stitched together (with some cross trace regularisation) to give an impedance section or volume. However, there is a significant difference between the seismic and EM cases. Whereas in the seismic case this approach is a fairly good approximation, in the case of marine CSEM (in particular) it is in general not: Because EM methods make a measurement of a volume of the earth the effect of lateral variations in resistivity within that volume cannot be easily ignored. Resistivity sections or volumes derived using a 1D approach should therefore be used only for looking at vertical and lateral bulk trends in resistivity and anisotropy and developing starting models for higher dimensional interpretation approaches.

Two-dimensional inversion approaches assume that the earth is invariant along one horizontal direction, usually taken to be perpendicular to the survey line (deGroot-Hedlin & Constable, 1990; Unsworth et al., 1993; MacGregor, 1999; Key and Ovall, 2011; Key, 2016). MT and CSEM data may be inverted either separately or jointly to give vertical resistivity sections. 2D inversion approaches also have the advantage of being computationally efficient, allowing rapid testing of data content, inversion parameterisations and constraints to determine the optimum approach and ensure the result is robust. Synthetic forward modelling, inversion and hypothesis testing can be used to further understand the inversion result. If full 3D inversion is required, the 2D stage is useful in ensuring a robust starting model, built from a number of 2D results.

Three-dimensional forward modelling and inversion approaches (see, e.g., Maao, 2007; Commer and Newman, 2008; Nguyen et al., 2016) provide the best approximation to the true resistivity structure and are now widely used in EM data interpretation. However, 3D inversion can be time consuming, particularly for CSEM, with run times on the order of days to reach a converged solution. For this reason, the construction of robust starting models using 1D and 2D approaches, and the constraint of the inversion itself with well log, seismic or geological information, is critical to ensure the 3D inversion process is as efficient as possible. The result is an isotropic or anisotropic resistivity volume.

In all cases it should be noted that the results of inversion, if unconstrained, provide a smoothed view of the resistivity structure of the earth. This is because marine EM methods are diffusive: The fields are not sensitive to the difference between a sharp change in resistivity and a more gradual change in properties.

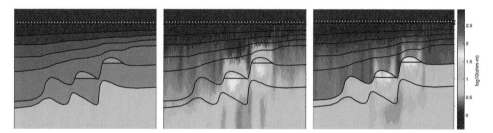

Figure 6.7 Illustration of constrained versus unconstrained inversion. (Left) input model from which synthetic data were generated. (Middle) An unconstrained inversion of the data gives a smoothed version of the input. (Right) Structural constraints from seismic data improve the resolution of the result. Results generated using the MARE2DEM code described by Key (2016).

This is illustrated with a synthetic example in Figure 6.7. The model in the left-hand panel was used to generate synthetic data, which was then contaminated with realistic survey noise. This synthetic dataset was inverted with no constraints to recover the resistivity section in the middle panel, which is a smoothed version of the input model. Resolution can be improved by including structural constraints from seismic data in the inversion process, imposing seismic scale resolution of boundaries onto the EM inversion (MacGregor and Sinha, 2000; Hansen and Mittet, 2009; Lovatini et al., 2009; Alvarez et al., 2017). This is illustrated in the right-hand panel of Figure 6.7. In this inversion, sharp jumps in resistivity are allowed in the inversion at known stratigraphic boundaries. The structure is now considerably better resolved and can be more easily interpreted alongside seismic or other data. Further examples are shown in the case studies later in this chapter.

In choosing an analysis approach, both technical considerations and constraints imposed by budget and timing must be taken into account. In some situations, a 2D inversion approach, carefully constrained with seismic structural information, can provide the information required to meet survey goals. Inclusion of information from other potential field methods can further constrain the response. While 3D inversion will provide a more complete and accurate model of the subsurface and is therefore often preferred, it can be time consuming compared to lower dimensional approaches. The workflow must be tailored to meet project requirements. The result of the analysis is an understanding of the distribution of resistivity in the earth. The resistivity sections/volumes resulting from interpretation must then be interpreted in terms of the underlying rock and fluid properties, and this requires careful integration of the results with seismic, well log or other potential field data in a multiphysics analysis.

6.7 Multiphysics Analysis: Combining Data Types to Understand the Earth

There are two main classes of geophysical problem where resistivity derived from marine EM methods can be used in conjunction with seismic or other data to improve understanding of the subsurface:

1. Structural problems where the goal is to understand the structure and stratigraphy of the subsurface. In cases in which seismic is unable to provide a clear image, for example, beneath basalt or salt layers, incorporation of resistivity information can assist in construction of an accurate migration velocity model, resulting in improved imaging of structure (see, e.g., MacGregor and Sinha, 2000; MacGregor, 2003; Key et al., 2006; Moorkamp et al., 2010, 2013, 2017; de Stephano et al., 2011; Colombo et al., 2012, 2013; Stadtler et al., 2014; Hoversten et al., 2015; Panzner et al., 2016).

2. Characterisation problems, where the goal is to understand the properties (e.g., lithology or fluid content) of regions of the earth. In cases where seismic alone cannot (or cannot reliably) predict properties in a reservoir or prospect, resistivity measurements, if carefully calibrated and interpreted, can substantially reduce the uncertainty in the result (see, e.g., Hoversten et al., 2006; Harris et al., 2009; Chen and Hoversten, 2012; Gao et al., 2012; MacGregor, 2012a; Alvarez et al., 2017, 2018; Du et al., 2017; Granli et al., 2018).

However, multiphysics analysis is not without challenges. First, different physical measurements must be coupled together through an earth model, that accurately and consistently describes each. This is typically achieved either through rock physics which relates each measurement to the underlying rock and fluid properties (e.g., Hoversten et al., 2006; Alvarez et al., 2017, 2018), or using structural coupling approaches (Gallardo and Meju, 2004, 2007; Moorkamp et al., 2013). Second, there must be an overlap in sensitivity between the methods applied in the zone of interest. Finally, the different scales at which measurements are made by each method must be reconciled.

There are many approaches to the combination of EM and other data types (MacGregor and Tomlinson, 2014), ranging from the qualitative to the fully quantitative. These include

- Qualitative approaches such as simple co-rendering of multiple data types to establish areas of correlation or lack thereof (Lovatini et al., 2009; MacGregor et al., 2012; Alcocer et al., 2013). Such approaches are almost universally applicable and provide a valuable first look at multiple data types. However, they take no account of the cause of the variations and correlations observed and so can be misleading.

- Integrated interpretation approaches where seismic and CSEM data are inverted separately in parallel workflows to yield elastic and electric properties respectively. These physical properties are linked through rock physics in order to understand the underlying cause of observed variations (Harris et al., 2009; MacGregor, 2012a; Morten et al., 2012; Alvarez et al., 2017, 2018). This is achieved through comparison of (for example) EM and seismic derived properties in a common domain (typically the electrical domain). Petrophysical joint inversion takes this further: Seismic and EM data are first inverted separately for elastic and electric properties respectively, which are then inverted jointly for the underlying rock and fluid properties (Andreis et al., 2018; Miotti et al., 2018). To apply methods such as this, the EM and seismic data must first be carefully analysed separately before the resulting physical properties are combined. Both seismic and CSEM data must demonstrate good sensitivity to the subsurface target and its properties. In general, well log data is required to calibrate the link between electric and other domains, making such approaches less applicable in frontier areas. However, where applicable, sequential quantitative integration workflows such as this provide an effective way of addressing the challenges of multiphysics analysis.
- Fully quantitative approaches such as joint inversion, where the different data types, including seismic, EM and potential field data, are coupled within the inversion algorithm. Such approaches have found success in structural imaging problems (de Stephano et al., 2011; Moorkamp et al., 2013; Heinke et al., 2014), resulting in improved images of salt or basalt. The problem is more complex if the object of the analysis is to determine reservoir rock or fluid properties. In addition to requirements for good sensitivity and well log–based calibration, the overburden structure must be well understood and ideally relatively benign in character. As a result, applications of joint inversion in this domain remain in the realm of research (Hoversten et al., 2006; Fleidner and Treitel, 2011; Chen and Hoversten, 2012; Gao et al., 2012; Liang et al., 2012).

There is seldom a one-size-fits-all approach to multiphysics analysis. Workflows must be tailored to address the specific challenge being studied. This will be illustrated using case studies.

6.8 Case Studies

This section presents two types of case studies: (1) improving seismic imaging using marine EM data and (2) an example of integrating a number of geophysical techniques to improve reservoir characterisation.

6.8.1 Improved Sub-basalt Imaging through the Integration of Seismic, CSEM and MT Data

There are a number of situations in which seismic struggles to provide an accurate representation of subsurface structure, for example, in the presence of salt or basalt which obscure deeper layers.

In the first example, the benefit of combining EM and seismic data when imaging potentially prospective sub-basalt sediments is considered. Basalt sequences present a challenge for seismic data because in general they have high acoustic impedance compared to the surrounding sediments and are also often interlayered with lower impedance material such as volcanoclastics and sediments. Whereas the top basalt reflector can be mapped with some accuracy, details beneath this are often obscured. In particular, the thickness of the basalt and the presence (or not) of sediments beneath can be hard to determine. There are examples where drilling has highlighted errors of more than 1 km in the seismic interpretation of base basalt (Hoversten et al., 2015). Given the cost of drilling in such environments, the motivation for improving the imaging and interpretation of the basalt and sediments beneath is clear.

Figure 6.8 shows a simplified 2D structure consisting of a layer of resistivity 200 Ωm which thickens from 1 km to 3 km representing a basalt flow, overlying a lower resistivity region representing a package of sub-basalt sediments with resistivity 10 Ωm. These are overlain by 1 km of sediments and underlain by a resistive crystalline basement. The structure is assumed to be isotropic in this simple example. The water depth is taken to be 1 km. There are three challenges to address:

1. Can the thickness of the basalt be resolved?
2. Can the presence and properties of the sub-basalt sediments be identified?
3. Can the depth to basement (or the thickness of the sediments) be determined?

To answer these questions, a synthetic CSEM and MT survey is undertaken. Five receivers are deployed on the seafloor (labelled R1 to R5 in Figure 6.8). For the CSEM survey these record multi-azimuth data from five source tow lines towed along the invariant direction of the 2D model. The transmission frequency is 1 Hz. The same five receivers also record MT data, in this case providing data at periods of 5–1,000 s. Although in both cases this represents a relatively limited survey layout compared to those used in practice, it serves to illustrate the relative strengths and weaknesses of the CSEM and MT data. Both data types were contaminated with Gaussian noise to mimic realistic survey conditions. The synthetic data were then inverted, both separately and jointly using the Occam algorithm of Constable et al. (1987), implemented as described in MacGregor

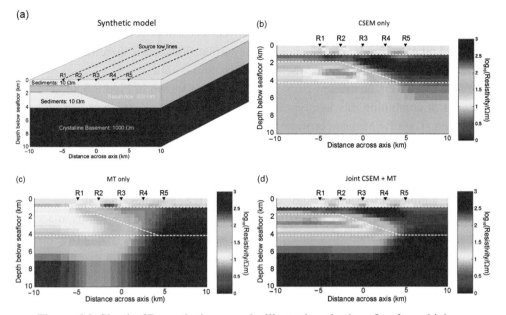

Figure 6.8 Simple 2D synthetic example illustrating the benefit of combining CSEM and MT data. (a) Synthetic model representing a basalt flow, sub-basalt sediments and basement. Synthetic CSEM and MT data were generated from this model, assuming five seafloor receivers (R1–R5) and five along strike source tows. (b) Inversion of the CSEM data alone. (c) Inversion of MT data alone. (d) Joint inversion of the CSEM and MT data. See text for discussion. (Example redrawn from MacGregor (2003))

and Sinha (2000) and extended to the joint inversion case in MacGregor (2003). In all cases it is assumed that the top of the basalt can be imaged effectively using seismic data. This boundary is therefore included as a break in the smoothness constraint in the inversion.

Looking first at the inversion of the CSEM data by itself (Figure 6.8b), it is clear that the resistive basalt layer is well resolved, especially where it is thin. The horizontal dipole used in the CSEM method applied in this example excites vertical current loops in the earth, making the resulting data particularly sensitive to resistive features that are thin compared to their depth of burial (MacGregor and Sinha, 2000; Constable and Srnka, 2007). The presence of the lower resistivity sub basalt sediments is also well resolved. However, looking at the deeper structure, the presence of the resistive basement, and hence the thickness of the sub-basalt sedimentary sequence is not clear. The CSEM signals do not penetrate far enough to resolve the structure at this depth.

Turning to the inversion of the MT data alone (Figure 6.8c), the greater depth penetration of the MT signals now allows the presence of the basement to be resolved. The sub-basalt sediments are also resolved as a low resistivity feature. However, in this case the basalt is not resolved where it is at its thinnest. The MT method relies on predominantly horizontal current flow and is therefore much less sensitive to resistive structure when such structure is thin.

The joint inversion of the CSEM and MT data is shown in Figure 6.8d, and the benefit of combining the data types is clear. The sub-basalt structure is now well resolved. The CSEM data constrain the basalt (especially where it is thin) and the sub-basalt sediments. The MT data provide further constraint on the sub-basalt sediments, but also the basement. Taken together CSEM and MT data provide a good image of the resistivity structure.

The simple 2D example shown in Figure 6.8 assumes that the basalt is homogeneous and isotropic. In reality, basalt is extremely heterogeneous, made up of stacks of individual flows interspersed by sediment, tuff and weathering layers. This makes the basalt extremely anisotropic, and this electrical anisotropy has a significant effect on the measured EM response.

To illustrate this, Figure 6.9a shows a 1D model in which the basalt layer is constructed from a stack of layers of varying thickness and resistivity. The distribution of layer thicknesses is based on mapped basalt flows in eastern Iceland (Smallwood et al., 1998), and layer resistivities increase with layer thickness. Layers are sampled randomly from the thickness/resistivity distribution and stacked to make the complete flow sequence to be modelled. This is representative of resistivity variations observed in basalt sequences in well log data (Pandey et al., 2008; Hoversten et al., 2015). The rapidly changing resistivity leads to a highly anisotropic basalt layer. Also shown in Figure 6.9a are the bulk harmonic average resistivity, representative of an effective horizontal resistivity (red) and the bulk arithmetic average resistivity, representative of an effective vertical resistivity (blue).

Figure 6.9b shows the amplitude and phase of the inline CSEM electric field, which most closely resembles the isotropic model corresponding to the vertical resistivity of the anisotropic model. Since the bulk vertical resistivity is relatively high in this case, CSEM sensitivity to the basalt is good. In contrast, the MT response shown in Figure 6.9c most closely resembles that of the bulk horizontal resistivity of the basalt layer. Not only is the MT response less sensitive to a resistive basalt layer in general if it is thin, heterogeneities in the basalt, leading to a lower horizontal resistivity may further erode the sensitivity. It is therefore important to take such electrical anisotropy into account.

Table 6.3 summarises the CSEM and MT sensitivity to basalt and sub-basalt structure in terms of the three questions posed at the start of this section.

Joint MT–CSEM analysis has been applied in practice to the problem of sub-basalt imaging: Hoversten et al. (2015) present an excellent case study on the

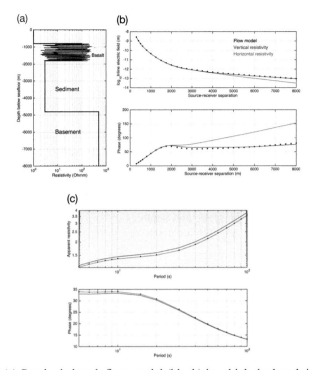

Figure 6.9 (a) Synthetic basalt flow model (black) in which the basalt is composed of flows of varying thickness and resistivity. This makes the resulting model extremely anisotropic. In this case the equivalent horizontal resistivity (bulk harmonic average) is 14 Ωm (red) and the equivalent vertical resistivity (bulk arithmetic average) is 102 Ωm (blue). (b) Amplitude and phase of the inline CSEM electric field. Black symbols show the response of the basalt flow model, while red and blue curves show the response of the equivalent horizontal and vertical resistivity models. (c) MT apparent resistivity and phase for the flow model (black symbols), equivalent horizontal (red) and vertical resistivity (blue) models.

inversion of MT and CSEM data to constrain basalt and sub-basalt structure in the Faroe–Shetland trough (Figure 6.10). Two wells in the area are used for calibration and validation of results. The top basalt boundary is assumed known in the inversion, and vertical transverse isotropy (VTI is assumed). In this calibration survey there is excellent agreement between the well log data, the seismic data and the results of joint CSEM and MT inversion. The basalt, sub-basalt sediments and basement are well resolved, demonstrating the ability of high-quality CSEM and MT data to improve seismic interpretation of base basalt and top basement (hence sediment thickness).

Multiphysics analysis in the context of structural interpretation has the most impact when it is used to update or improve seismic interpretation (MacGregor, 2012b; Panzer et al., 2016). In both cases marine CSEM and MT data were used to

Table 6.3 *Summary of the relative sensitivity of CSEM and MT methods*

	MT	CSEM
Can the thickness of the basalt be resolved?	MT has low sensitivity to resistive basalt layers unless they are thick compared to their burial depth.	CSEM is extremely sensitive to thin (compared to burial depth) basalt sequences.
Can the presence and properties of the sub-basalt sediments be identified?	MT is sensitive to the bulk properties of the sub-basalt sediments (but not details of the structure).	CSEM is sensitive to the bulk properties of the sub-basalt sediments (but not details of the structure).
Can the depth to basement (or the thickness of the sediments) be determined?	MT is sensitive to the presence and depth of resistive basement.	CSEM will not constrain basement if the depth is greater than 3–4 km.

guide the construction of a velocity model to improve the seismic imaging of sediments obscured by a layer of basalt. Initial seismic imaging provides structural cofnstraints, primarily the depth to the top of the basalt layer. This is used as a constraint in the inversion of the CSEM and MT data (separately and/or jointly) for an initial resistivity volume. From this, using an appropriate link between velocity and resistivity (either empirically derived from well logs or based on rock physics), a new velocity model can be constructed and the seismic data re-imaged. This process is repeated until a good result is achieved. Results make clear that that the integration of marine EM–derived resistivity information dramatically improves the quality of the seismic image in the sub-basalt area.

This approach has the advantage that it can be undertaken using widely available tools and algorithms without the need for specialist technology. Intermediate results can also provide insight into the problem, and the workflow can be updated if necessary. This process can of course be encapsulated within a joint inversion algorithm, and excellent results have been obtained (see, e.g., Jegen et al., 2009).

6.8.2 Multiphysics Imaging of Sub-salt Structure: Combining EM, Gravity and Seismic Information

The challenge of imaging around salt bodies can also be addressed using a multiphysics approach, combining seismic with marine EM and other technologies. Accurate characterisation of salt bodies and the surrounding strata using seismic alone can be difficult, because of often low signal-to-noise ratio seismic, large velocity contrasts between salt and surroundings and complex salt geometries.

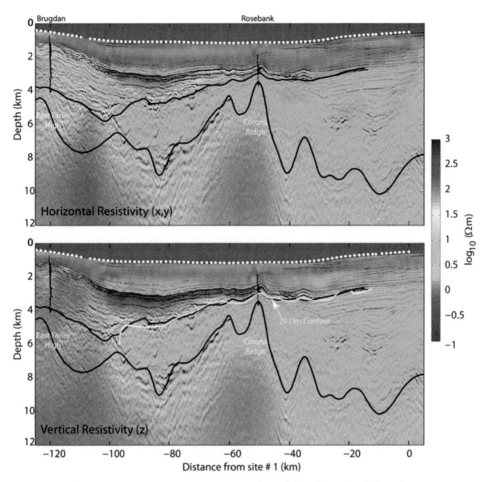

Figure 6.10 Example showing the constraint of basalt and sub-basalt structure using the joint inversion of CSEM and MT data. Thick black lines show base basalt (upper line) and top basement (lower line) interpolated from 2D seismic lines. The white line in the vertical resistivity section shows the 20 Ωm contour which agrees well with the seismically defined boundary. (From Hoversten et al., 2015. © 2015 European Association of Geoscientists and Engineers)

Early marine MT surveys in the Gulf of Mexico focussed on determining the geometry of salt bodies (Key et al., 2006). However, determining the geometry of the salt using resistivity is only the first step. The ultimate goal is to improve the seismic image obtained, through the combination of seismic and EM methods. Medina et al. (2012) combined marine MT and seismic data from the Walker Ridge area of the Gulf of Mexico in a simultaneous joint inversion for resistivity and velocity. The resulting models were used to update interpretations of the base of salt, and the geometry of the feeders between the allochthonous salt bodies and deeper autochthonous salt. The updated velocity model was then used in a reverse time migration which gave a significantly improved image of salt and sub-salt structure.

Colombo et al. (2018) combined marine CSEM and MT with seismic data from the Red Sea, where halite bodies are embedded within layered evaporate sequences. Reconstruction of velocity in this complex overburden setting is a challenge. A hierarchical workflow was applied to improve the velocity model and hence the seismic image ultimately obtained. In the first stage of the workflow, CSEM and MT data were inverted for a detailed 3D resistivity distribution. Seismic travel times were used to derive a robust velocity starting model. The seismic data were then inverted using a structural coupling to the horizontal resistivity model derived from the CSEM inversion (which showed the best consistency with the known geology of the area) to obtain a new and geologically more consistent velocity model. Finally, the seismic data were migrated using a Kirchhoff pre-stack depth migration and the new velocity model. Details of the layered evaporite sequence were much better imaged, the base of the evaporites was clearer and more continuous, and hints of the basement were observed.

It is not only the combination of seismic and marine EM data that can be valuable. Moorkamp et al. (2013) demonstrate the use of seismic, MT and gravity data to constrain the structure of a marine salt dome. A structurally coupled joint inversion of these three contrasting data types is applied, and the resulting velocity and resistivity values are compared to well logs in the area. Figure 6.11 shows the results of the inversion of the three data types separately (top row) and together (bottom row). The salt dome is observed as a high-velocity, high-resistivity, low-density feature in the subsurface. When inverted in isolation, the three models return very different structures, with the salt as imaged by the MT being considerably more extensive than in either the velocity or gravity inversion results. When inverted jointly using a structure-coupled approach, the three property models are far more consistent. This is particularly clear in the density result, where the structural coupling between data types has resulted in a deeper and more extensive density anomaly compared to that obtained from the gravity data alone.

As well as structurally more consistent models, the properties themselves are also more accurately recovered by the joint inversion. This is illustrated in Figure 6.12, where the velocity and resistivity from the inversion are compared to values derived from boreholes in the flanks of the salt diapir. When the data are inverted separately, the velocity and resistivity of the background sediments and relationship between these are relatively well recovered. However, only the joint inversion correctly recovers the properties of the salt itself.

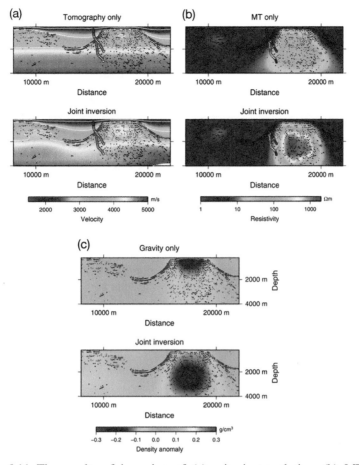

Figure 6.11 The results of inversion of (a) seismic travel time (b) MT and (c) gravity data alone (top row) and using a structure-coupled joint inversion (bottom row). (From Moorkamp et al., 2013 © 2013 American Geophysical Union, all rights reserved)

6.8.3 *Improved Reservoir Characterisation through Integrated Interpretation of CSEM and Seismic Data*

Reservoir characterisation involves the determination of subsurface lithology and fluid properties and also benefits from multiphysics analysis using seismic and CSEM data. These benefits have been demonstrated in the Hoop area of the Barents Sea. Exploration success has been mixed in this area: The discovery of the Wisting field in 2013 (well 7324/8-1), comprising 60 m of oil charged sand in a Jurassic shore face deposit approximately 300 m below seafloor, was followed by an oil discovery at Hansen (7324/7-2) and a small gas discovery at Mercury (7324/9-1) in the same formation. All of these cases showed strong seismic indications of hydrocarbon charge. However, the seismic indications were not infallible.

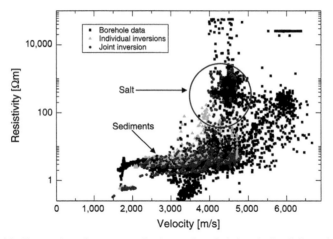

Figure 6.12 Comparison between velocity and resistivity derived from individual (green) and joint inversion (purple) of seismic, MT and gravity data, compared to properties derived from boreholes in the flanks of the salt diapir (black). The joint inversion recovers the velocity–resistivity relationship both within the salt and in the surrounding sediments. (From Moorkamp et al., 2013. © 2013 American Geophysical Union, all rights reserved)

Following these successes, Apollo (7324/2-1) and Bjaaland (7324/8-2) were drilled on similar seismic indications and discovered only residual hydrocarbon charge. Based on these results it became clear that exploration in this area could not rely solely on seismic data. This provided the impetus for a number of multiphysics studies (Alvarez et al., 2017, 2018; Granli et al., 2018).

The limitation of the seismic analysis is explained in Figure 6.13, which shows cross plots of elastic and electric attributes from the wells in the area (Alvarez et al., 2018). Acoustic impedance is plotted against Poisson's ratio in Figure 6.13a for the background trend (in grey) and for four fluid cases within the reservoir (colours). There is some separation between the wet and hydrocarbon charged cases; however, there is little to distinguish commercial hydrocarbon saturations from residual (fizz) gas. The resulting seismic responses are near identical. Electrical resistivity is included in Figure 6.13b. In a multiphysics cross plot space the ambiguity is resolved and commercial hydrocarbon saturation can be clearly identified. Note, however, that it is not possible to distinguish between a commercial saturation of oil and gas: Both lead to high resistivity.

Alvarez et al. (2017) used multiphysics analysis of seismic and CSEM data to make just such a distinction. The data comprised six lines of 2D seismic data and towed streamer CSEM data acquired concurrently. Two wells were used for calibration purposes: Wisting Central (7324/8-1), a discovery and Wisting alternative (7324/7-1S) which encountered only water-bearing sands in the

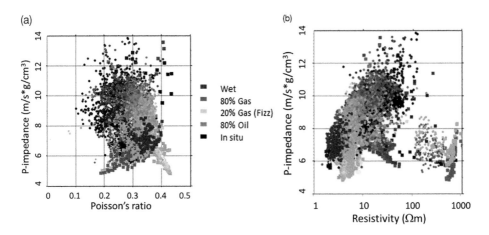

Figure 6.13 (a) Cross plot of elastic properties from wells in the Hoop area. Using elastic data alone it is virtually impossible to distinguish commercial hydrocarbon saturations from residual (fizz) gas. This ambiguity has led to the drilling of a number of non-commercial wells. (b) Incorporating electric attributes resolves the ambiguity. In a multiphysics cross plot domain, commercial and residual saturations are clearly distinct. (From Alvarez et al., 2018)

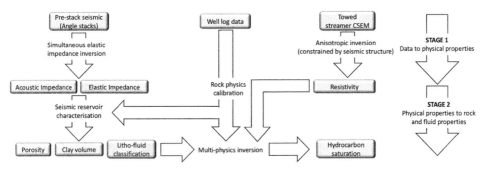

Figure 6.14 The multiphysics workflow proceeds in two stages. During stage 1, geophysical data are inverted for physical properties, in this case the acoustic and elastic impedance, and the electrical resistivity. In stage 2 these attributes are used to determine the underlying rock and fluid properties. Rock physics, calibrated using well log data, provides the link between the electric and elastic domains.

Jurassic reservoir zone of interest. These data were used to predict two wells that were drilled subsequently: Hanssen (7324/7-2), a discovery, and Bjaaland (7324/8-2) that encountered sub-commercial residual hydrocarbons.

The multiphysics workflow applied is illustrated in Figure 6.14. In stage 1, data were inverted for physical properties. A simultaneous elastic impedance inversion was used to invert seismic angle stacks for acoustic and elastic impedance. In a parallel workflow, CSEM data were inverted for anisotropic resistivity, using seismic horizons to constrain the structure and ensure consistency between seismic and EM results (a pre-requisite for integration). In stage 2, the physical properties were used to determine the underlying rock and fluid properties. A seismic reservoir characterisation workflow, calibrated at the wells, was used to derive porosity, volume of clay and litho-fluid facies. Rock physics can then be used to link the electric and elastic domains. In this example the Simandoux relationship was applied (Simandoux, 1963), again calibrated using the well log data. This links the measured bulk resistivity (derived from CSEM data) to the porosity and clay content (derived from the seismic data) and the fluid saturation, the parameter to be determined. Using the seismic and CSEM derived attributes as input, and the Simandoux equation as the link between electric and elastic domains, it is then possible to invert for the reservoir saturation, using the seismically derived litho-fluid facies as a geological framework.

Figure 6.15 shows the result of the multiphysics analysis along a section passing through the Wisting Central, alternative and Hanssen locations, and close to the Bjaaland well location. The upper panel in Figure 6.15 shows the seismically derived litho-fluid facies. Using seismic data alone, the Wisting Central discovery

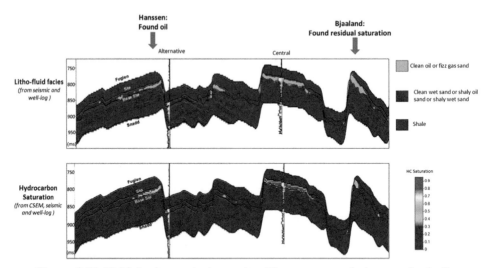

Figure 6.15 Multiphysics analysis results. The upper panel shows seismically derived litho-fluid facies. Both Hanssen and Bjaaland are predicted to contain hydrocarbon-bearing sands. The lower panel shows the hydrocarbon saturation derived from multiphysics analysis, which predicts a significant hydrocarbon charge at Hanssen but only residual saturation at Bjaaland. These results were confirmed with drilling. (Re-drawn from Alvarez et al., 2017)

is clearly resolved, and the reservoir interval is wet at the Alternative location. Hanssen and Bjaaland look similar on the basis of the seismic reservoir characterisation, both with a high probability of being hydrocarbon charged.

The lower panel in Figure 6.15 shows the hydrocarbon saturation in the reservoir interval resulting from the multiphysics analysis. In contrast to the seismic-only result, Hanssen and Bjaaland now look significantly different. At Hanssen there is clear indication of a significant hydrocarbon charge. This was confirmed when the well was drilled. However, at Bjaaland, where the seismic indicated the presence of hydrocarbon charged sands, the hydrocarbon saturation is low and confined to the very crest of the structure. When drilled, the Bjaaland well encountered only residual (sub-commercial) hydrocarbon confirming the results of the multiphysics analysis. More detail on this example can be found in Alvarez et al. (2017).

Alvarez et al. (2018) show a second example from the Hoop area of the Barents Sea. The goal of this study was to use 3D seismic and nodal CSEM data to understand the prospectivity of the area, with a view to making a decision on whether the block had sufficient commercial promise to warrant drilling, or whether the block should be relinquished. A similar workflow to that shown in Figure 6.14 was used to obtain physical properties: acoustic and elastic impedance from the seismic and anisotropic resistivity from the CSEM. However, in this case a multiphysics attribute analysis was used to integrate these measurements and understand the underlying rock and fluid properties (Figure 6.16).

Figure 6.16a shows a multiphysics cross plot space. The *y*-axis shows the probability of finding hydrocarbon saturated sand based on seismic data alone. The CSEM-derived resistivity is shown on the *x*-axis. Using seismic data alone there is an ambiguity between residual and commercial hydrocarbon saturations. Similarly, if we consider only CSEM-derived resistivity there is an ambiguity between commercial hydrocarbon saturations, and other potentially resistive geological features such as carbonates or tight sands. Data from the reservoir interval are projected into this space in Figure 6.16b, colour coded by point density. The majority of points plot in the lower left quadrant of the area, representing water saturated sands as would be expected.

Taking cut-offs based calibration at the well log in the area (Apollo, which encountered residual hydrocarbon), a multiphysics litho-fluid facies map of the area was created (Figure 6.16c). Towards the north-west of the area, it is clear that there are more resistive lithologies present (red colours in Figure 6.16c). This corresponds to a trend of decreasing porosity in this direction, identified in both the seismic and CSEM data. The prospect of interest in this area is outlined in white. The multiphysics analysis makes clear that the probability of finding commercial hydrocarbon is low. Taken together with other exploration considerations in the area, the license partnership agreed to relinquish the license. This of course means

(a)

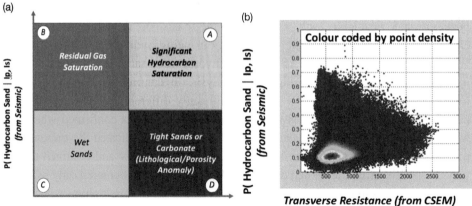

(b)

(c)

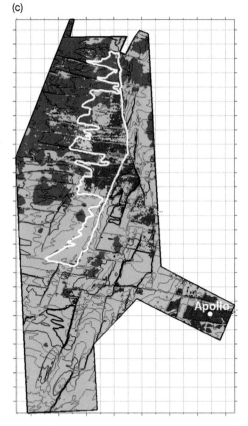

Figure 6.16 (a) Multiphysics cross plot space in which ambiguities in both the seismic and EM data, when considered alone, can be resolved. (b) Data from the reservoir interval colour coded by point density, plotting in this space. Most points lie in the lower left quadrant, indicating water-saturated sand. (c) Multiphysics litho-fluid classification. The prospect of interest is outlined in white and is likely to contain only residual hydrocarbon saturation. The well used for calibration, Apollo (7324/2-1), is marked.

that, for the moment at least, we do not have drilling confirmation of the result. However, it highlights two important points. First, a negative result can be just as important and valuable in an exploration setting as a positive one. Second, analysis such as this is valuable only if the results, positive or negative, lead to action.

6.9 Use in the Exploration and Exploitation Workflow

Marine EM can be applied throughout the exploration and exploitation workflow. However, risks and uncertainties vary (MacGregor and Cooper, 2010). The interpretation uncertainty falls as more complementary information, for example seismic and well log data, is available. Table 6.4 summarises the applications, risks and uncertainties at each stage.

Table 6.4 *Summary of applications of marine EM through the oil field life cycle*

	Complementary data available.	MT	CSEM
Frontier exploration/ regional play analysis	No wells/ little or no seismic. Some gravity or magnetic.	Mapping large scale background resistivity structure/depth to basement across a basin. Salt/basalt mapping.	CSEM as a complement to MT in areas of thin salt/basalt. CSEM can map resistivity variations over a wide area, but interpretation risk in terms of underlying properties is extremely high in the absence of calibration.
Exploration – creation of lead inventory	Some wells, sparse seismic	Detailed mapping of salt/basalt structure. Background/basement structure to constrain CSEM analysis.	CSEM interpretation risk is moderate and case dependent: depends on availability of background information and calibration.
Prospect evaluation/ Appraisal	3D seismic and wells	Background/basement structure to constrain CSEM analysis.	CSEM interpretation is focused on hydrocarbon extent: a well constrained and calibrated problem.
Field development and monitoring	3D seismic, wells and production data	n/a	CSEM interpretation is focused on changes in fluid extent and/or properties: a well constrained and calibrated problem.

6.9.1 Frontier Exploration/Regional Play Analysis

In frontier exploration there is unlikely to be well log data available for calibration, and perhaps only sparse (or no) seismic data. Marine EM methods can be applied effectively in two ways: First, MT may be applied for mapping large scale background resistivity variations, depth to basement and features such as salt and/or basalt. Regional MT surveying of this sort can provide a useful precursor to more detailed seismic or CSEM studies. In some cases, regional CSEM can provide additional constraint on thin salt/basalt structures (those less than about 1 km thick) and on gross scale trends in resistivity and resistivity anisotropy in the background structure. Examples of this application include Colombo et al. (2013) in the Red Sea, Ceci et al. (2014) in the Gulf of Mexico and Heincke et al. (2014) and Hoversten et al. (2015) in the west of Shetland area.

CSEM has also been applied in frontier areas as a means of assessing prospectivity. In some cases this has been successful (e.g., Lovatini et al., 2009; Fanavoll et al., 2012); however, it should be borne in mind that the interpretation risk is high. Although CSEM in a frontier area can map resistivity variations, without constraints or calibration from seismic or well log data the interpretation of these variations is likely to be non-unique and ambiguous. The poor structural resolution of the CSEM method means that the resulting resistivity images are diffuse, and the uncertainty in the depth of features is large, so that they may not be unambiguously attributed to a particular stratum. If there are multiple resistive features, these cannot be easily separated, and small resistive bodies are likely to be lost or smoothed into surrounding strata.

Even assuming that localised resistivity anomalies can be found, the cause of these anomalies cannot be unambiguously linked to the presence of hydrocarbon. Is the high resistivity zone observed related to a fluid effect or to a carbonate or tight sandstone? This question cannot be answered without calibration, which is unlikely to be available in a frontier area. In a number of cases an uncalibrated resistive feature has been drilled (at some expense); however, the cause was found to be lithological rather than fluid related (Gist et al., 2013). This is not a failure of the CSEM method: Resistivity was correctly mapped. It is failure to link the resistivity feature to the underlying subsurface property. Resistivity, taken alone, cannot be relied on as an indicator of hydrocarbon charge. Similarly the absence of resistivity variations, although decreasing the likelihood, does not necessarily preclude the presence of commercial hydrocarbons: Survey design deficiencies as a result of poorly understood background geology may result in a survey failing to detect charged structures.

6.9.2 Targeted Exploration and Prospect Ranking

By far the majority of marine EM surveys to date fall into this application category, for which seismic data and well data are likely to be available to complement the marine EM. Since the focus of the marine EM survey is the potential reservoir, CSEM methods play the greatest role, although MT still provides useful (and in some cases critical) information on background and basement structure.

The challenge to be addressed is as follows: Given a seismically defined structure and information on lithology from well log data, can CSEM provide information to de-risk the likely fluid content? Seismic data provide information on the reservoir structure (but potentially not in its content or extent) and seismic and well log information provide information on the surrounding strata in which the reservoir is embedded. The CSEM interpretation problem is therefore much better constrained, and the resulting interpretation risk is consequently lower.

Examples of this application include Alcocer et al. (2013) in the Gulf of Mexico, MacGregor et al. (2012) offshore West Africa and Nordskag et al. (2013) and Alvarez et al. (2018) in the Barents Sea. Buland et al. (2011) and Hesthammer et al. (2012) provide a useful overview of the success rates of this approach. Statistics quoted by Shell suggest a discovery success rate of 80% through the integration of CSEM surveys into the exploration workflow (compared to approximately 47% before the integration of CSEM), with a 100% success in the prediction of dry holes pre-drill (Karman et al., 2011). However, the choice of targets to which CSEM is applied by Shell is made carefully and has been predominantly in Tertiary settings. Price et al. (2019) present a benchmarking study which at least approximately supports these high success rate. Similarly, another operator in the Asia Pacific region has reported that of ten wells drilled on high resistivity anomalies derived from CSEM surveys, six were discoveries, with the remaining four penetrating high-resistivity lithology that was not hydrocarbon bearing, again highlighting the importance of careful interpretation of CSEM results to mitigate lithology risk.

6.9.3 Appraisal

Reservoir appraisal is a still better constrained challenge for CSEM, that seeks to address the question of the extent of a known hydrocarbon accumulation away from well control, given knowledge of the seismic structure and properties. MT is less applicable, except in cases where detailed basement mapping can complement the CSEM results (for example, if the target reservoir lies close to resistive electrical basement).

A good example of this is the appraisal survey performed in 2009 on the Kraken field in the UK North Sea. The survey was designed to map the extent of the known hydrocarbon accumulation, which was not possible using seismic data alone. The results of the CSEM analysis supported further successful drilling to appraise the reservoir (Nautical Petroleum Annual Report, 2010). Further examples of this application include appraisal of a North Sea chalk reservoir (MacGregor, 2012a), appraisal of the Troll field in the Norwegian North Sea (Morten et al., 2012) and examples in the Barents sea including Alvarez et al. (2017) and Granli et al. (2018).

6.9.4 Reservoir Monitoring

By the time a field is producing it is well characterised: Its structure is known and the fluid distribution has been estimated and modelled. The challenge to be addressed is to map changes in the fluid content and distribution over time. CSEM-derived resistivity variations should be well placed to address this question, given tight structural constraints from seismic and reservoir models. Changes in gas content of an onshore storage reservoir have been mapped using land-based EM (Ziolkowski et al., 2007). For the offshore case, a large number of modelling studies have been undertaken (Lien and Mannseth, 2008; Orange et al., 2009; Andreis and MacGregor, 2011; Lin et al., 2012) which indicate that under certain conditions successful 4D reservoir monitoring using CSEM is feasible; however, this application has yet to be tried in practice.

Note that since MT methods are largely insensitive to the thin resistive reservoir structures, the MT method is not applicable in this setting.

Table 6.5 provides a summary of the uses of marine EM in the context of the other technologies discussed in this book.

6.10 Indicative Acquisition Costs

At the time of writing, marine EM acquisition is far from a commodity, with only two commercial contractors and various academic groups collecting data. The cost depends strongly on the survey type, objective and geographic location making it hard to generalise. For surveys in which seafloor receivers are utilised (nodal CSEM, vertical dipole CSEM and MT), The cost of marine CSEM acquisition depends on two main factors:

- Spacing between receivers. Typical surveys use between a 1×1 km and 3×3 km grid. Note that the grid need not be regular and can be designed to optimise sensitivity to the targets of interest.

Table 6.5 *Summary of marine EM uses in the context of other technologies described in this book*

E&P Phase	Basin screening	Access	Exploration	Prospect evaluation	Appraisal	Development	Production
Crustal geophysics							
Gravity and magnetics							
Full tensor gravity							
Marine CSEM						Not tested	Not tested
Marine MT							
Ocean bottom seismic							
Micro-seismic and passive seismic							
Conventional reflection seismic							

CSEM, controlled-source electromagnetic; E&P, exploration and production; MT, magnetotelluric. Green indicates a good fit; orange, partial fit; blank, not applicable.

- The number of tow lines or source positions used. For marine CSEM with a towed horizontal dipole, it is usual to tow the source over each line of receivers deployed in one direction. Tie lines in the orthogonal direction can be useful on occasion, and it is also possible to increase coverage by towing the source between the lines of receivers. For vertical dipole surveys, where the source is stationary during transmission it is usual to collect source stations between the receivers.

To cover a given survey area with a 1×1 km receiver grid (and associated tow lines) is considerably more costly than covering the same area with a 3×3 km receiver grid over which the source is towed. The choice between these options needs to be driven both by survey design considerations ensuring that the survey objectives are met, and budgetary constraints. Similarly, the cost scales with the total area to be covered: prospect specific surveys with a tailored survey geometry will be less costly than large regional surveys.

The towed streamer CSEM system allows more efficient acquisition of CSEM data. No direct cost comparisons are yet available; however, anecdotal evidence suggests that the cost may be between two and three times lower than the equivalent node-based coverage. Note that this efficiency comes at the expense of multi-azimuth coverage, and therefore the choice of acquisition systems should be based on survey optimisation for the target and problem to be addressed in addition to budgetary considerations.

The acquisition of marine MT data is generally charged per site: the more MT sites required, the more costly the survey. Note that most node based CSEM receivers can record MT during a CSEM survey deployment: In this instance the cost of the MT data, acquired alongside CSEM is significantly reduced.

6.11 Future Trends

Marine EM methods have had a somewhat chequered history over the past 20 years. Initial excitement prompted over-marketing and under-delivery, and the resulting disappointment led to a dramatic reduction in market size. Recently, with a more cautious and realistic approach being applied, the popularity of CSEM and MT methods is once again growing. Surveys are becoming more efficient and cost effective, further driving uptake of the technology.

The focus here is on applications of marine EM applications in the hydrocarbon industry. However, it is worth noting that such methods are also applicable to the detection and characterisation of other resources. Monitoring CO_2 in offshore carbon capture and storage projects is an interesting application of CSEM methods, potentially providing early warning of CO_2 leakage. Modelling studies have been undertaken to demonstrate the efficacy of the method (Bhuyian et al., 2012); however, there has been no field validation of results. There has been recent increase in interest in gas hydrate mapping, one of the earliest marine EM applications (see Weitemeyer et al., 2011; Bedanta et al., 2016 or Attias et al., 2017 for more recent examples). Shallow marine EM applications have been applied to mapping freshwater aquifers around coastlines (see, e.g., Blatter et al., 2019; Gustafson et al., 2019; Lippert and Tezkan, 2020). Finally marine EM methods have been applied to the mapping of submarine massive sulphide deposits (e.g., Constable et al., 2017; Gehrmann et al., 2019), which have the potential to become an important economic resource to satisfy the worldwide growth in demand for metals.

6.12 Summary

Marine EM is a robust tool for measuring the resistivity of the subsurface. If applied and interpreted carefully the results can provide valuable complementary information to that obtained from other geophysical methods. Some general principles apply to ensuring successful CSEM surveys.

- CSEM and MT methods can be applied to measure resistivity over a large area for the purpose of understanding basin geology and context, in combination with other remote sensing methods.
- Should the survey objective concern prospectivity of an area, ensure good background information is available. CSEM, in particular, is most valuable in areas where well log and seismic data are available. Interpretation risk is high in frontier areas. MT may be applied in these situations to map background, basement and salt/basalt structures.
- Choose the targets sensibly, and plan surveys carefully. For any marine EM exercise the first stage is to understand the problem to be addressed and best approach to use. CSEM in particular is not a general tool and is not applicable in all environments. Careful screening is required to find those areas where CSEM will add the most value.
- Do not interpret CSEM or MT derived resistivity estimates in isolation. Resistivity information in itself can be ambiguous and careful integration with seismic, well log or other complementary data is required to understand the underlying lithology and fluid drivers responsible for the resistivity variations observed. This point applies equally to all remote sensing data.

If applied and interpreted carefully the results can provide valuable information complementary to that obtained from other geophysical methods.

6.13 Parameters and Units

Table 6.6 *Table of parameters and units relevant to EM methods*

Parameter	Description	Unit
B	Magnetic flux density	T (Tesla)
D	Electric displacement	C/m^2 (Coulombs/metre^2)
E	Electric field. Often presented normalised to unit source dipole moment.	V/m or V/Am^2 when normalised to unit SDM
ε	Electric permittivity – the typical free space value $\varepsilon_0 = 8.85 \times 10^{-12}$ F/m. Only affects high frequency EM methods (kHz or more).	F/m (Farads/metre)
H	Magnetic field	A/m
J	Current density. $J = \sigma E$ where σ is conductivity and E is the electric field.	A/m^2
μ	Magnetic permeability. In free space this has the value $4\pi \times 10^{-7}$. Can typically be assumed to take its free space value in EM problems.	H/m (Henris/metre) or alternatively Newtons/Ampere^2
ρ	Resistivity	Ωm (Ohm metre)
σ	Conductivity, the inverse of resistivity	S/m (Siemens/metre)
SDM	Source dipole moment: the measure of transmitter strength.	Am (Ampere metre)

References

Alcocer, J. A. E, Garcia, M. V., Soto, H. S., et al., 2013. Reducing uncertainty by integrating 3D CSEM in the Mexican deepwater workflow. *First Break,* **31**, 75–9.

Alumbaugh, D., Cuevas, N., Chen, J., Gao, G. and Brady, G., 2010. Comparison of sensitivity and resolution with two marine CSEM exploration methods. *Expanded Abstract, SEG Annual Meeting, Denver 2010.*

Alvarez, P., Alvarez, A., MacGregor, L., Bolivar, F., Keirstead, R. and Martin, R., 2017. Reservoir properties prediction integrating controlled source electromagnetic, pre-stack seismic and well log data using a rock physics framework: Case study in the Hoop Area, Barents Sea, Norway. *Interpretation,* **5**(2), SE43–SE60.

Alvarez, A., Marcy, F., Vrijlandt, M., et al., 2018. Multi-physics characterisation of reservoir prospects in the Hoop area of the Barents Sea, *Interpretation,* **6**(3), SG1–SG17.

Andreis, D. and MacGregor, L, 2011. Using CSEM to monitor production from a complex 3D gas reservoir: A synthetic case study. *The Leading Edge,* September, 1070–9.

Andreis, D., MacGregor, L., Grana, D., Alvarez, P. and Ellis, M. 2018. Overcoming scale incompatibility in petrophysical joint inversion of surface seismic and CSEM data. *Expanded Abstract, SEG Annual Meeting.*

Attias, E., Weitemeyer, K., Holz, S., et al., 2017. High resolution imaging of marine gas hydrate structures by combined inversion of CSEM towed and ocean bottom receiver data. *Geophysical Journal International,* **214**, 1701–14.

Barker, N. D., Morten, J. P. and Shantsev, D. V., 2012. Optimizing EM data acquisition for continental shelf exploration. *The Leading Edge, November 2012, 1276–84.*

Bedanta, K. G., Weitemeyer, K., Minshull, T., Sinha, M., Westbrook, G. and Marin-Moreno, H., 2016. Resistivity image beneath an area of active methane seeps in the west Svalbard continental slope. *Geophysical Journal International, 207,* 1286–302.

Bhuyian, A. H., Landro, M. and Johansen, S. E., 2012. 3D CSEM modelling and time-lapse sensitivity analysis for sub-surface CO_2 storage, *Geophysics,* **77**(5), E343–E355.

Blatter, D., Key, K., Ray, A., Gustafson, C. and Evans, R., 2019. Bayesian joint inversion of controlled source electromagnetic and magnetotelluric data to image freshwater aquifer offshore New Jersey. *Geophysical Journal International,* **218**, 1822–37.

Bouchrara, S., MacGregor, L., Alvarez, A., et al., 2015. CSEM based anisotropy trends across the Barents Sea. *Expanded Abstract, SEG Annual Meeting,* New Orleans, 2015. http://dx.doi.org/10.1190/segam2015–5912741.1.

Buland, A., Loseth, L. O., Becht, A., Roudot, M. and Rosten, T, 2011. The value of CSEM data in exploration. *First Break,* **29**, 69–76.

Ceci, F., Clementi, M., Guerra, I. and Mantovani, M., 2014. Integrated interpretation and simultaneous joint inversion of CSEM and seismic datasets – The sunshine case, *Expanded Abstract, EAGE Annual Meeting, 2014.*

Chave, A., Constable, S. and Edwards, R. N., 1991. Electrical exploration methods for the seafloor. In M. N. Nambighian (ed.), *Electromagnetic Methods in Applied Geophysics,* 931–96. Tulsa, OK: Society of Exploration Geophysicists.

Chen, J. and Hoversten, G. M., 2012. Joint inversion of marine seismic AVA and CSEM data using statistical rock physics and Markov random fields, *Geophysics,* **77**, R65–R80.

Colombo, D., Keho, T. and McNeice, G., 2012. Integrated seismic-electromagnetic workflow for sub-basalt exploration in northwest Saudi Arabia. *The Leading Edge,* January, 42–52.

Colombo, D., MacNiece, G., Curiel, E. S. and Fox, A., 2013. Full tensor CSEM and MT for subsalt structural imaging in the Red Sea: Implications for seismic and electromagnetic integration. *The Leading Edge,* April, 436–49.

Colombo, D., Rovetta, D. and Turkoglu, E., 2018. CSEM regularised seismic verlocity inversion: A multiscale, hierarchical workflow for sub-salt imaging. *Geophysics,* **83**, B241–B252.

Colombo, D. and Stefano, M. D., 2007. Geophysical modeling via simultaneous joint inversion of seismic, gravity, and electromagnetic data: Application to prestack depth imaging. *The Leading Edge, 26*(3), 326–31.

Commer, M. and Newman, G., 2008. New advances in three dimensional controlled source electromagnetic inversion. *Geophysical Journal International,* **172**, 513–35.

Constable, S., 2010. Ten years of marine CSEM for hydrocarbon exploration. *Geophysics,* **75**, A67–A81.

Constable, S., 2013. Instrumentation for marine magnetotelluric and controlled source electromagnetic sounding. *Geophysical Propsecting,* **61**, 505–32.

Constable, S. and Cox, C. S., 1996. Marine controlled source electromagnetic sounding II: The PEGASUS experiment. *Journal of Geophysical Research,* **101**, 5519–30.

Constable, S. C., Kowalczyk, P. and Bloomer, S., 2017. Measuring marine self potential using an autonomous underwater vehicle. *Geophysical Journal International,* **200**, 1–8.

Constable, S., Orange, A., Hoversten, M. and Morrison, H. F., 1998. Marine magnetotellurics for petroleum exploration, Part 1: A sea floor equipment system. *Geophysics*,
63, 816–25.

Constable, S., Parker, R. and Constable, C., 1987. Occam's inversion: A practical algorithm for generating smooth models from electromagnetic sounding data.
Geophysics, **52**, 289–300.

Constable, S. and Srnka, L., 2007. An introduction to marine controlled source electromagnetic methods for hydrocarbon exploration, *Geophysics*, **72**, WA3–WA12.

Cox, C. S., 1981. On the electrical conductivity of the oceanic lithosphere. *Physics of the
Earth and Planetary Interiors*, **25**(3), 196–201.

Cuevas, N. H. and Alumbaugh, D., 2011. Near source response of a resistive layer to a
horizontal or vertical electric dipole excitiation. *Geophysics*, **76**, F353–F371.

Davydycheva, S. and Frenkl, M. A., 2013. The impact of 3D tilted resistivity anisotropy on
CSEM measurements. The Leading Edge, 32, 1374–81.

Du, Z., Namo, G., May, J., Reiser, C. and Midgley, J., 2017. Total hydrocarbon volume in
place: Improved reservoir characterisation from integration of towed streamer EM
and dual sensor broadband seismic data. *First Break*, **35**, 89–96.

Edwards, R. N., 1997. On the resource evaluation of marine gas hydrate deposits using
seafloor transient dipole-dipole measurements. *Geophysics*, **62**, 63–74.

Edwards, R. N., 2005. Marine controlled source electromagnetics: Principles, methodologies and future commercial applications. *Surveys in Geophysics*, **26**, 675–700.

Ellingsrud, S., Eidesmo, T., Johansen, S., Sinha, M. C., MacGregor, L. M. and Constable,
S., 2002. Remote sensing of hydrocarbon layers using sea-bed logging (SBL):
Results of a cruise offshore West Africa. *The Leading Edge*, **21**, 972–82.

Ellis, M., MacGregor, L., Vera de Newton, P., et al., 2017. Investigating electrical anisotropy drivers across the Barents Sea. *Expanded Abstract, EAGE Annual Meeting,
Paris, 2017.*

Ellis, M., Ruiz, F., Nanduri, S., et al., 2011. Impotance of anisotropic rock physics
modelling in integrated seismic and CSEM interpretation. *First Break*, **29**, 87–95.

Ellis, M. E., Sinha, M. C. and Parr, R., 2010. Role of fine scale layering and grain
alignment in the electrical anisotropy of marine sediments. *First Break*, **28**, 49–56.

Engelmark, F., Mattsson, J., McKay, A. and Du, Z., 2014. Towed streamer EM comes of
age. *First Break*, **32**, 75–8.

Evans, R. L., Sinha, M. C., Constable, S. and Unsworth, M. J., 1994. On the electrical
nature of the axial melt zone at 13° North on the East Pacific Rise. *Journal of
Geophysical Research*, **99**, 577–88.

Fanavoll, S., Ellingsrud, S., Gabrielsen, P. T., Tharimela, R. and Ridyard, D., 2012.
Exploration with the use of EM data in the Barents Sea: The potential and the
challenges. *First Break*, **30**, 89–96.

Fliedner, M. and Treitel, S. 2011. Stochastic inversion of CSEM and seismic data using the
Neighbourhood Algorithm. *Extended Abstracts, EAGE Annual Meeting.*

Francis, T. J. G., 1977, Electrical prospecting on the continental shelf. *British Geological
Survey Report*, 77–4.

Gabrielsen, P. T., Abrahamson, P., Panzer, M., Fanavoll, S. and Ellingsrud, E., 2013.
Exploring frontier areas using 2D seismic and 3D CSEM data, as exemplified by
multi-client data over the Skrugard and Havis discoveries in the Barents Sea. *First
Break*, **31**, 63–71.

Gallardo, L. A. and Meju M. A., 2004. Joint two-dimensional DC resistivity and seismic
travel time inversion with cross-gradients constraints. *Journal of Geophysical
Research*, **109**, B03311. http://dx.doi.org/10.1029/2003JB002716.

Gallardo, L. A. and Meju M. A., 2007. Joint two-dimensional cross-gradient imaging of magnetotelluric and seismic traveltime data for structural and lithological classification. *Geophysical Journal International*, **169**, 1261–72, http://dx.doi.org/10.1111/j.1365–246X.2007.03366.x.

Gao, G., Abubakar, A. and Habashy, T. M., 2012. Joint petrophysical inversion of electromagnetic and full waveform seismic data. *Geophysics*, **77**, D53–D68.

Gehrmann, R., North, L. J., Graber, S., et al., 2019. Marine mineral exploration with controlled-source electromagnetics at the TAG hydrothermal field, 26N Mid-Atlantic Ridge. *Geophysical Research Letters*. http://dx.doi.org/10.1029/2019GL082928.

Gist, G., Ciucivara, A., Houck, R., Rainwater, M., Willen, D. and Zhou, J-J., 2013. Case study of a CSEM false positive. *Expanded Abstract, SEG Annual Meeting, Houston, 2013,* http://dx.doi.org/10.1190/segam2013–0307.1.

Granli, J. R., Daudina, D., Robertson, S. C., Morten, J. P., Gabrielsen, P. and Sigvathsen, B., 2018. Applying high-resolution 3D CSEM and seismic for integrated reservoir characterization. *Expanded Abstract, 88th SEG Annual Meeting*, 2018, 949–53, https://doi.org/10.1190/segam2018–2995351.1.

de Groot-Hedlin, C. and Constable, S., 1990. Occam's inversion to generate smooth, two-dimensional models from magnetotelluric data. *Geophysics*, **55**, 1613–24.

Gustafson, C., Key, K. and Evans, R., 2019. Aquifer systems extending far offshore on the US Atlantic margin. *Scientific Reports*, **9(1)**. https://doi.org/10.1038/s41598-019-44611-7.

Hansen, K. R. and Mittet, R., 2009. Incorporating seismic horizons in inversion of CSEM data. *SEG Technical Program Expanded Abstracts*, **2009**, 694–8.

Hansen, K., Panzer, M., Shantsev, D. and Mittet, R., 2016. TTI inversion of CSEM data. *Expanded Abstract, 86th EAGE Annual Meeting*.

Hansen, K., Panzer, M., Shantsev, D. and Mohn, K., 2018. Comparison of TTI and VTI 3D inversion of CSEM data. *Expanded Abstract, EAGE Annual Meeting, Copenhagen, 2018*.

Harris, P., Du, Z., MacGregor, L., Olsen, W., Shu, R. and Cooper, R., 2009. Joint interpretation of seismic and CSEM data using well log constraints: An example from the Luva field. *First Break*, **27**, 76–81.

Heincke, B., Moorkamp, M., Jegen, M., Hobbs, R. W. and Berndt, C., 2014. 2D and 3D joint inversion of seismic, MT and gravity data from the Faroe Shetland Basin. *Expanded Abstract, EAGE Annual Meeting 2014*.

Helwig, S., Wahab El Kaffas, A., Holten, T., Frafjord, O. and Eide, K., 2013. Vertical dipole CSEM: Technology advances and results from the Snovhit field. *First Break*, **31**, 63–8.

Helwig, S., Wood, W. and Gloux, B., 2019. Vertical-vertical controlled source electromagnetic instrumentation and acquisition. *Geophysical Prospecting*, **67**, 1582–94.

Hesthammer, J., Stefatos, A. and Sperrevik, S., 2012. CSEM efficiency – evaluation of recent drilling results. *First Break*, **30**, 47–55.

Hokstad, K., Fotland, B., Mackenzie, G., et al., 2011. Joint imaging of geophysical data: Case history from the Nordkapp Basin, Barents Sea. *Extended Abstract, SEG Annual Meeting 2011*, 18–23.

Holten, T., Flekkoy, E. G., Singer, B., Blixt, E. M., Hanssen, A., and Maloy, K. J., 2009. Vertical source, vertical receiver electromagnetic technique for offshore hydrocarbon exploration. *First Break*, **27**, 89–93.

Hoversten, G. M., Cassassuce, F., Gasperikova, E., et al., 2006. Direct reservoir parameter estimation using joint inversion of marine seismic AVA and CSEM data. *Geophysics*, **71**, C1–C13.

Hoversten, G. M., Constable, S. and Morrison, H. F., 2000. Marine magnetotellurics for base-of-salt mapping: Gulf of Mexico field test at the Gemini structure. *Geophysics*, **65**, 1476–88.

Hoversten, G. M., Morrison, F. and Constable, S. C., 1998. Marine magentotellurics for petroleum exploration, part II: Numerical analysis of sub-salt resolution, *Geophysics*, **63**, 826–40.

Hoversten, G. M., Myer, D., Key, K., Alumbaugh, D., Hermann, O. and Hobbet, R., 2015. Field test of sub-basalt hydrocarbon exploration with marine controlled source and magnetotellutic data. *Geophyical Prospecting*, **63**, 1284–310.

Jegen, M. D., Hobbs, R. W., Tarits, P. and Chave, A., 2009. Joint inversion of marine magnetotelluric and gravity data incorporating seismic constraints. Preliminary results of sub-basalt imaging off the Faroe Shelf. *Earth and Planetary Science Letters*, **282**(1–4), 47–55.

Karman, G., Ramirez, D., Voon, J. and Rosenquist, M., 2011. A decade of controlled-source electromagnetic, CSEM, in Shell: Lessons from a global look back study. *Presented at the 4th NPF Biennial Petroleum Geology Conference, Bergen.*

Karpiah, A. B., Meju, M., Miller, R. and Musafarudin, R., 2019. Improving basement depth mapping using 3D marine magnetotelluric (MT) inversion. *Expanded Abstract, SEG Annual Meeting 2019.* http://dx.doi.org/10.1190/segam2019-3214939.1.

Key, K., 2009. 1D inversion of multicomponent, multifrequency marine CSEM data: Methodology and synthetic studies for resolving thin resistive layers. *Geophysics*, **74**, F9–F20.

Key, K., 2011. Marine electromagnetic studies of seafloor resources and tectonics. *Surveys in Geophysics.* http://dx.doi.org/10.1007/s10712-011-9139-x.

Key, K., 2016. MARE2DEM: A 2D inversion code for controlled source electromagnetic and magnetotelluric data. *Geophysical Journal International*, **207**, 571–88.

Key, K., Constable, S. and Weiss, C., 2006. Mapping 3D salt using the 2D marine magnetotelluric method: Case study from Gemini Prospect, Gulf of Mexico, *Geophysics*, **71**, B17–B27.

Key, K. and Ovall, J., 2011. A parallel goal-oriented adaptive finite element method for 2.5–D electromagnetic modelling. *Geophysical Journal International*, **186**, 137–54.

Klein, J. D., Martin, P. R. and Allen, D. F., 1997. The petrophysics of electrically anisotropic reservoirs. *The Log Analyst*, **38**, 25–36.

Liang, L., Abubaker, A. and Habashy, T., 2012. Joint inversion of controlled source electromagnetic and production data for reservoir monitoring. *Geophysics*, **77**, ID9–ID22.

Lien, M. and Mannseth, T., 2008. Sensitivity study of marine CSEM data for reservoir production monitoring. *Geophysics*, **73**, F151–F163.

Lin, L., Abubaker, A. and Habashy, T. M., 2012. Joint inversion of controlled-source electromagnetic and production data for reservoir monitoring. *Geophysics*, **77**, ID9–ID22.

Lippert, K. and Tezkan, B., 2020. On the exploration of a marine aquifer offshore Israel by long offset transient electromagnetics. *Geophysical Prospecting*, **68**, 999–1015.

Lovatini, A., Umpbach, K. and Patmore, S., 2009. 3D CSEM in a frontier basin offshore Greenland. *First Break*, **27**, 95–8.

Maao, F., 2007. Fast finite-difference time-domain modelling for marine-subsurface electromagnetic problems. *Geophysics*, **72**, A19–A23.

MacGregor, L., 2003. Joint analysis of marine active and passive source EM data for sub-salt or sub-basalt imaging. *Expanded Abstract, 65th EAGE Annual Conference, Stavanger, 2003.* https://doi.org/10.3997/2214-4609-pdb.6.F18.

MacGregor, L. and Cooper, R. C., 2010. Unlocking the value of CSEM. *First Break,* **28,** 49–54.

MacGregor, L. M., 1999. Marine controlled source electromagnetic sounding: Development of a regularised inversion for 2D resistivity structures. *LITHOS Science Report*, **1**, 103–9.

MacGregor, L. M., 2012a. Integrating seismic, CSEM and well log data for reservoir characterization. *The Leading Edge,* March, 268–77.

MacGregor, L. M., 2012b. Integrating seismic, CSEM and well log data for characterisation of reservoirs. *EAGE Workshop on EM in Hydrocarbon Exploration, April 2012, Singapore.*

MacGregor, L. M., Bouchrara, S., Tomlinson, J. T., et al., 2012. Integrated analysis of CSEM, seismic and well log date for prospect appraisal: A case study from West Africa. *First Break*, **30**, 77–82.

MacGregor, L. M., Constable, S. C., and Sinha, M. C., 1998. The RAMESSES experiment III: Controlled source electromagnetic sounding of the Reykjanes Ridge at 57°45'N. *Geophysical Journal International*, **135**, 772–89.

MacGregor, L. M. and Sinha, M. C., 2000. Use of marine controlled source electromagnetic sounding for sub-basalt exploration. *Geophysical Prospecting*, **48**, 1091–106.

MacGregor, L. M., Sinha, M. C. and Constable, S., 2001. Electrical resistivity structure of the Valu Fa Ridge, Lau Basin, from marine controlled source electromagnetic sounding. *Geophysical Journal International,* **146**, 217–36.

MacGregor, L. and Tomlinson, J., 2014. Marine controlled source electromagnetic methods in the hydrocarbon industry: A tutorial on method and practice, *Interpretation*, **2**, AH13–SH32.

MacGregor, L., Tomlinson, J. and Maver, K. G., 2019. CSEM acquisition methods in a multi-physics context. *First Break,* **37**, 67–72.

Mattsson, J., Englemark, F. and Anderson, C., 2013. Towed streamer EM: The challenges of sensitivity and anisotropy. *First Break,* **31**, 155–9.

Mattsson, J., Lindqvist, P., Juhasz, R. and Bjornemo, E., 2012. Noise reduction and error analysis for a towed EM system. *Expanded Abstracts, SEG Technical Program 2012,* 1–5.

McKay, A., Mattsson, J. and Du, Z., 2015. Towed streamer EM – reliable recovery of subsurface resistivity. *First Break*, **33**, 75–85.

Medina, E., Lovatini, A., Andreasi, F. G., Re, S. and Snyder, F., 2012. Simultaneous joint inversion of 3D seismic and magnetotelluric data from the walker ridge. *First Break*, **30**, 85–8.

Miotti, F., Zerilli, A., Menezes, P., Crepaldi, J. and Vianna, A., 2018. New petrophysical joint inversion workflow: Advancing on reservoir characterisation challenges, *Interpretation*, **6**, SG33–SG39.

Mittet, R. and Schaug-Pettersen, T., 2008. Shaping optimal transmitter waveforms for marine CSEM surveys, *Geophysics,* **73**, F97–F104.

Moorkamp, M., 2017. Integrating electromagnetic data with other geophysical observations for enhanced imaging of the earth: A tutorial and review. *Surveys in Geophysics,* **38**, 935–62.

Moorkamp, M., Heincke, H., Jegen, M., Roberts, A. W. and Hobbs, R. W., 2010. A framework for 3D joint inversion of MT, gravity and seismic refraction data. *Geophysical Journal International*, **184**, 477–93.

Moorkamp, M., Roberts, A. W., Jegen, M., Heincke, B. and Hobbs, R. W., 2013. Verification of velocity-resistivity relationships derived from structural joint inversion. *Geophysical Research Letters*, **40**, 1–6.

Morten, J. P., Roth, F., Carlsen, S. A., et al., 2012. Field appraisal and accurate resource estimation from 3D quantitative interpretation of seismic and CSEM data. *The Leading Edge*, **31**, 447–56.

Myer, D., Constable, S. and Key, K., 2010. Broad band waveforms and robust processing for marine CSEM surveys. *Geophysical Journal International*, **184**, 689–98.

Newman, G., Commer, M. and Carazzone, J., 2010. Imaging CSEM data in the presence of anisotropy. *Geophysics*, **75**, F51–F61.

Nguyen, A. K., Hanssen, P., Mittet, R., et al., 2017. The next generation electromagnetic acquisition system. *Expanded Abstract, 79th EAGE Annual Conference and Exhibition, Paris, 2017.*

Nguyen, A. K., Nordskag, J. I., Wiik, T., et al., 2016. Comparing large scale 3D Gauss-Newton and BFGS CSEM inversions. *Expanded Abstract, SEG 86th Annual Meeting, 2016.*

Nordskag, J., Kjosnes, A, Nguyen, K. and Hokstad, K., 2013. CSEM exploration in the Barents Sea: Joint seismic and CSEM interpretation. *Expanded Abstract, SEG Annual Meeting 2013.*

Orange, A. S., 1989. Magnetotelluric exploration for hydrocarbons. *Proceedings of the IEEE*, **77**, 287–317.

Orange, A., Key, K. and Constable, S., 2009. The feasibility of reservoir monitoring using time-lapse marine CSEM. *Geophysics*, **74**, F21 –F29

Pandey, D., MacGregor, L., Sinha, M. and Singh, S., 2008. Feasibility of using magneto-telluric for sub-basalt imaging at Kutch, India. Applied Geophysics, **5**(1), 74-82.

Panzner, M., Morten, J. P., Weilull, W. W. and Borge, A., 2016. Integrated seismic and electromagnetic model building applied to improve subbasalt depth imaging in the Faroe-Shetland Basin. *Geophysics*, **81**(1), E57–E68.

Price, A., Twarz, C. and Gabrielsen, P., 2019. Building confidence in CSEM for exploration – Benchmarking. *Expanded Abstract, 89th SEG Annual meeting, San Antonio.* 10.1190/segam2019–3214720.1.

Ramananjaona, C. and MacGregor, L., 2010. 2.5D inversion of CSEM data in a vertically anisotropic earth. *Journal of Physics, Conference Series,* **255**, 012004.

Ramananjaona, C., MacGregor, L. and Andreis, D., 2011. Inversion of marine electromagnetic data in a uniaxial anisotropic stratified earth. *Geophysical Prospecting*, **59**, 341–60.

Schlumberger, C., Schlumberger, M. and Leonardon, E. G., 1934. Electrical exploration of water-covered areas. *Transactions of the American Institute of Mining and Metallurgical Engineers*, **110**, 122–34.

Schwalenberg, K., Willoughby, E., Mir, R. and Edwards, R. N., 2005. Marine gas hydrate electromagnetic signatures in Cascadia and their correlation with seismic blank zones. *First Break*, **23**, 57–63.

Simandoux, P., 1963. Dielectric measurements in porous media and application to shaly formation. *Revue de L'Institut Français du Pétrole*, **18**, 193–215.

Sinha, M. C., Constable, S. C., Peirce, C., et al., 1998. Magmatic processes at slow spreading ridges: Implications of the RAMESSES experiment at 57°45' North on the Mid-Atlantic Ridge. **135**, 731–74.

Sinha, M. C., Patel, P. D., Unsworth, M. J., Owen, T. R. E. and MacCormack, M. R. J., 1990. An active source EM sounding system for marine use. *Marine Geophysical Research,* **12**, 59–68.

Smallwood, J., White, R. S. and Staples, R., 1998. Deep crustal reflectors under Reydarfjordur, eastern Iceland: Crustal accretion above the Iceland mantle plume. *Geophysical Journal International*, **134**, 277–90.

Srnka, L. J., 1986. Method and apparatus for offshore electromagnetic sounding utilizing wavelength effects to determine optimum source and detector positions. U. S. Patent 4,617,518.

Stadtler, C, Fichler, C., Hokstad, K., Myrlund, E. A., Wienecke, S. and Fitland, B., 2014. Improved salt imaging in a basin context by high resolution potential field data: Nordkapp Basin, Barents Sea. *Geophysical Prospecting,* **62**, 615–30.

de Stefano, M., Andreasi, F. G., Re, S., Virgilio, M. and Snyder, F. F., 2011. Multiple domain, simultaneous joint inversion of geophysical data with application to sub-salt imaging, *Geophysics,* **76**, R69–R80.

Tseng, H-W., Stalnaker, J., MacGregor, L. and Ackermann, R., 2015. Multi-dimensional analysis of the SEAM controlled source electromagnetic data: The story of a blind test of interpretation workflows. *Geophysical Prospecting*, **63**, 1383–402.

Unsworth, M., Travis, B. and Chave, D., 1993. Electromagnetic induction by a finite electric dipole over a 2D earth. *Geophysics*, **58**, 198–214.

Vozoff, K., 1972. The magnetotelluric method in the exploration of sedimentary basins. *Geophysics*, **37**, 98–141.

Weitemeyer, K., Constable, S. and Trehu, A., 2011. A marine electromagnetic survey to detect gas hydrate at Hydrate Ridge, Oregon. *Geophysical Journal International*, **187** (1), 45–62.

Young, P. D. and Cox, C. S., 1981. Electromagnetic active source sounding near the East Pacific Rise. *Geophysical Reseach Letters,* **8**, 1043–6.

Yuan, J. and Edwards, R. N., 2000. The assessment of marine gas hydrates through electrical remote sounding: Hydrate without the BSR? *Geophysical Research Letters,* **27**, 2397–400.

Ziolkowski, A., Hobbs, B. A. and Wright, D., 2007. Multitransient electromagnetic demonstration survey in France. *Geophysics*, **72**, 197–209.

7

Ocean Bottom Marine Seismic Methods

IAN JACK

7.1 Introduction

Why the trend towards ocean bottom seismic?

Ocean bottom seismic delivers the highest quality marine seismic data. It has geometrical flexibility so it allows unrestricted source/receiver geometry, and it can provide data close to or under platforms. Significant recent advances in technology have led to major reductions in cost. Consequently, ocean bottom seismic is on a strong upwards trend. It remains more expensive than many towed cable seismic configurations, but it is difficult to challenge its positive economics. Even for a small but typical recent find containing, say 25 mm bbl, the cost of the high-quality seismic data needed for exploration and development will be less than $1/bbl. 'Conventional' narrow azimuth towed cable seismic data, although cheaper, may neither find the reservoir nor contribute usefully to its development.

This chapter of the book gives a brief review of the development of 'conventional' marine seismic work from 2D through 3D, its shortcomings, and its continuing development into 'broadband' seismic. It then provides a comprehensive description of ocean bottom work and compares the costs of all these techniques. The focus of very recent developments in marine seismic has been on autonomous underwater vehicles, the use of simultaneous sources (or 'blending'), and the extension of the technology into much longer source–receiver distances combined with the generation of an extra octave of low frequencies to assist full-waveform inversion (FWI) in velocity model building.

The market share of ocean bottom work in the marine seismic business increased from 10% in 2013 to an estimated 25% in 2020.

7.2 History

Since its introduction in the late 1960s, most offshore seismic work has been conducted using towed detector cables and airgun sources. Simple 2D acquisition

Figure 7.1 A towed cable vessel of the 1970–1980 timeframe. (Courtesy of WesternGeco)

with a single cable and one or two sources is cheap per linear kilometre, and this configuration continues to see some use, mainly for reconnaissance exploration work in unexplored areas. Figure 7.1 is representative of such a vessel.

The move to 3D technology from the late 1980s resulted in vessels which could tow a dozen or more cables (the current maximum is 24 but this high number is not in general use), thus collecting wide swathes of data giving 3D images at low unit costs. Figure 7.2 shows such a vessel.

These 'conventional' 3D towed cable techniques allow construction of '3D seismic cubes' such as the one shown in Figure 7.3. In areas where the seismic data quality is good to fair as in this example, the 3D cube allows subsequent successful mapping of large simple structures.

However, where data quality is more challenging, improved resolution of the seismic data is needed, and in many situations conventional towed cable seismic

Figure 7.2 PGS Ramform Victory towing 16 streamers. (Courtesy of PGS)

Figure 7.3 Multi-client data. (Reprinted with permission from Geophysical Pursuit, Inc. & Fairfield Geotechnologies)

data are unable to provide the quality of imaging necessary for exploration and development work. Obvious examples include mapping top/base salt in North Sea environments, and subsalt developments in the Gulf of Mexico. Improvement in seismic resolution and imaging involves wider temporal bandwidth as well as improved source–receiver acquisition geometry.

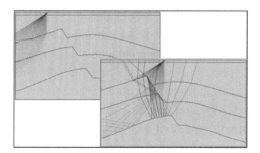

Figure 7.4 Ray-path scattering at geological boundaries.

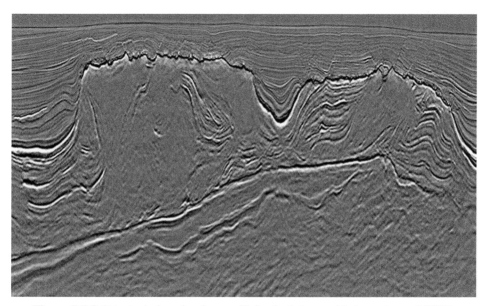

Figure 7.5 Salt tectonic imaging issues. (Seismic data provided courtesy of CGG)

Complex and heavily faulted geology is normal. In these situations ray paths are highly divergent in three dimensions, requiring much more extensive acquisition geometries. The example in Figure 7.4 shows what happens to source–receiver ray paths as a vessel approaches complexity (and only 2D scattering is shown here).

Salt tectonics (as in Figure 7.5) is an example of a geological environment in which successful imaging requires the best possible data acquisition and processing techniques. In practice this means good sampling in several domains including source–receiver offset and source–receiver azimuth.

In the past, towed cable techniques have struggled to fulfil these geometrical and bandwidth requirements. However, solutions have evolved which significantly improve the resolution of conventional towed cable seismic data, although they do so at a cost. In this context, the term broadband seismic' is now well established, and usually refers to one or more of these advanced towed streamer techniques, which will be described later. The percentage of marine seismic work recorded under the 'broadband' description has risen from single figures in 2007 to more than 50% in 2020 and continues to increase. These advances in marine recording took place mainly from 2007 and continue to the present. The techniques include contractor-specific terminology such as IsoMetrix^{TM}, GeoStreamer^{TM}, GeoSource^{TM}, Broadseis^{TM}, Stagseis^{TM}, Orion^{TM}, FreeCable^{TM} and SoundSabre^{TM}.

Other terminology in common use includes narrow-azimuth (NAZ); multi-azimuth (MAZ); wide-azimuth (WAZ); rich-, full- or all-azimuth (RAZ/FAZ/AAZ) and wide-angle towed streamer (WATS) data. These acronyms are described later. Most imply some 'stretching' of the geometry of conventional narrow azimuth towed cable configurations.

Early ocean bottom recording was acquired sparsely in a 2D mode by research departments, using autonomous retrievable ('pop-up') seismometer units (OBS). For exploration work, re-deployable cables (OBC) using electrical sensors and wired data transmission were available from the 1970s, and fibre-optic sensors and transmission from around 2000. Cables could be trenched into the seabed and connected to installations for permanent reservoir monitoring (PRM), sometimes also referred to as 'life of field seismic systems (LoFS)'.

Meanwhile, autonomous node (OBN) technology had been developing strongly. Nodes can be deployed in several different ways – they can be attached to wire or kevlar ropes and laid from a moving vessel, or they can be laid by ROV. There is currently much interest in the use of autonomous underwater vehicles (AUVs) either as the nodes themselves or as service vehicles for deploying nodes, retrieving their data, or charging their batteries.

The integration of downhole well data via optical fibres in wells with surface and ocean bottom seismic data is a growth area.

The term 'blended sources', which became commonplace on land surveys by 2010, began to appear on marine surveys from around 2015 and is now offered by several contractors. It can be applied to all of the towed cable and ocean bottom techniques and is discussed later.

A section view of a modern towed streamer configuration with 12 streamers and 3 sources in blended or unblended mode could look like Figure 7.6.

7.3 Marine Seismic Options and Survey Costs

Towed streamer 2D by its nature gives an inaccurate, poorly sampled and noisy picture of the subsurface, especially where the geology is dipping or complex. It sits at the bottom of a 'quality ladder' as illustrated in Figure 7.7. However, it is

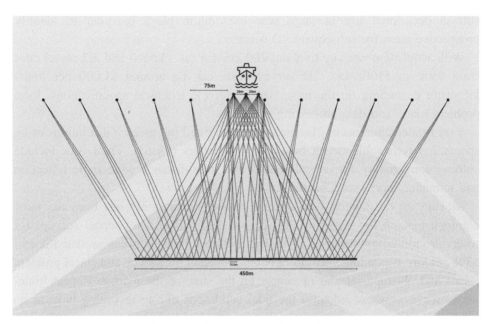

Figure 7.6 Blending of sources. (Courtesy of Shearwater)

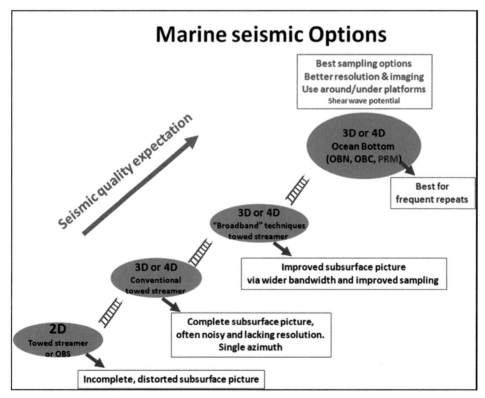

Figure 7.7 Quality expectations for marine seismic options.

still in occasional use in large new exploration plays in order to identify prospective areas for subsequent 3D surveys.

With acquisition rates up to about 200 km/day (at 5 knots) and 2D vessel rates from \$50k to \$100k/day, 2D survey costs can be around \$1,000 per linear kilometre depending on the area, size of survey, technical specifications, local problem issues and data processing.

Very limited amounts of 2D data are still acquired by research institutions using specialised retrievable ocean bottom seismometers (OBSs). Their uses include velocity tomography for crustal studies and long-distance wide-angle reflection and refraction work; see Chapter 3.

Moving up the ladder, conventional towed streamer 3D has been the most common method for acquiring 3D seismic data and also for repeat surveys for reservoir monitoring purposes (4D). The unit cost varies between around \$8k and \$30k per km^2 depending very strongly on the size of the survey area, local problem issues and obstructions and of course on the state of the market. Repeat 'time-lapse' surveys, where (ideally) the tidal conditions during recording have to be carefully matched to an earlier survey, are at the higher end of the price scale. The limitations of these surveys will be discussed shortly. Several techniques, referred to earlier as broadband', can be applied to towed streamer surveys which remove or reduce some of these limitations, albeit with substantial increases in cost. These surveys start from around \$15 k/km^2 and can easily be five times the cost of conventional towed cable 3D.

At the top end of the ladder, placing detectors on the seabed allows most if not all of the technical limitations of towed cable work to be overcome. A huge inventory of results is now available, most of which are compellingly good.

For OBN work, the unit costs increase to between \$30k and \$200k per km^2 depending strongly on the water depths and on the specific technique and parameters used. These costs initially limited the take-up of this type of work, but operational developments have taken place in the last few years which have allowed, and continue to allow, significant cost reductions, approaching the lower of these cost figures, with a resulting increase (10% to 25% from 2013 to 2020) in the OBN market share of marine seismic work.

As mentioned earlier, cabled systems (with electrical or fibre-optic sensors) can be entrenched for permanent reservoir monitoring purposes (PRM systems), although the upfront costs of these have again limited their deployment. They tend to be installed and cost effective where frequent repeat surveys are likely to be required (e.g., once or more per year) over several years and where many wells can be positioned using the results. The installation costs are of the same order as the equipment itself, and a total installed cost of around \$1 million per km^2 can be anticipated – but with relatively cheap repeat surveys.

A rising expectation of the need for continuing improvements in seismic quality, coupled with the large inventory of high-quality results from ocean bottom surveys, is driving the trend towards increased use of nodes. There is an acknowledgement that the cost increments can be quickly recouped by savings on wells, improved production, improved recovery and better reservoir management generally.

7.4 Conventional Towed Streamer Marine Seismic Acquisition (and Its Limitations)

Towed cable marine seismic data are collected mainly using hydrophone (pressure) detector sensors along straight lines. 2D surveys usually consist of a grid of widely spaced orthogonal lines, several kilometres apart, each separately processed. '3D' surveys normally use parallel lines along the direction of the length or width of the survey area and are processed as an entity. Single cables were standard from the 1960s until the early 1980s, after which the number of cables towed simultaneously has increased gradually to between 8 and 24. This has allowed fewer vessel passes giving cheaper unit costs on 3D surveys and also denser lateral sampling resulting in improved imaging. Streamer construction has also changed over the years, with oil-filled designs being replaced with 'solid' construction (mainly for environmental reasons) and with a greater linear density of hydrophone sensors. Conventional marine seismic data continue to be acquired this way. However, there are several problems and limitations inherent in the basic technique:

1. The hydrophone detectors receive first the upcoming wavefield, and then a few milliseconds later, the same wavefield with reversed polarity reflected downwards by the free surface. This results in serious temporal bandwidth reductions, illustrated in Figure 7.8.
2. Tides and currents cause the cables to 'feather' from side to side of the planned track, thus perturbing the consistency and regularity of the subsurface illumination between adjacent passes of the vessel, often requiring expensive 'infill' to be acquired on 3D surveys.
3. Since the airgun sources and detector cables are all attached to the vessel, there are geometry limitations (for a single vessel operation) in terms of the maximum source-to-detector distances which can be acquired. And in general, the more cables which are towed, the shorter they have to be. Only the most powerful vessels can tow more than a dozen cables each of 10 km length.
4. Another geometrical restriction is that the parallel swathes result in seismic coverage at near-constant source-to-receiver azimuth. Typical maximum

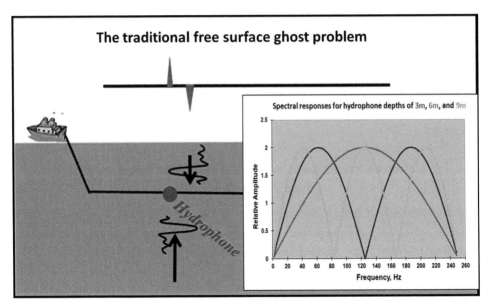

Figure 7.8 Bandwidth effects of the free surface ghost.

spreadwidths are around 1,500 m (e.g., 16 cables at a cable separation of 100 m). Even with the size and power of the latest seismic vessels it is difficult to envisage any significant increase in this dimension. However, considerable potential exists for imaging improvements if the azimuth domain can be populated; see Section 7.5. Small surveys are very inefficient, since the time taken for the vessel to turn 180° from one swathe to the next can be longer than the time taken to record a swathe.

5. Infrastructure such as platforms and floating production storage and offloading units (FPSOs) severely restrict the access of a vessel towing a swathe of detector cables, typically by a 500 m exclusion zone either side of the obstruction. In such cases, a second source vessel is usually employed to sail on opposite sides of the infrastructure to record a 'platform undershoot' to give reflection coverage beneath the obstruction. This reduces the imaging problems (but only to some extent) and increases the survey cost significantly. Platform undershoots usually present additional technical problems for time-lapse ('4D') surveys.

6. Repeat surveys for time-lapse ('4D') purposes require accurate replication of source and receiver positions. This is difficult in typical tidal environments, requiring careful pre-planning of tides and currents and 'streamer steering', which is effective only up to a limited angle of correction. It significantly increases the cost of repeat '4D' surveys.

7. Towing cables through the water imparts a certain amount of ambient noise to the detectors. This worsens with sea-state and the survey eventually has to be suspended until the sea-state subsides.
8. Shear-wave data cannot be acquired by towed cables.
9. The downgoing inverted polarity wavefield is recorded after the upcoming wavefield with a delay of $2d/V$, where d is the depth of tow and V is the velocity of sound in water. This imparts a severe spectral response to the data as a function of cable depth, as shown here for cables at three different depths. (Similar issues are incurred at the seismic source.)

7.5 'Broadband': Improving the Quality of Marine Seismic Data

Achieving 'broadband' marine seismic data has two major components. First, it implies the use of method(s) which restore the temporal bandwidth notches illustrated in Figure 7.8 which are inherent in the conventional towed cable seismic method. Second, it involves reductions in the levels of several types of noise.

Progress in either of these components generally requires care and innovation both in data acquisition and in data processing, and these two activities are intimately linked. Data processing improvements often follow developments in acquisition such as improved spatial sampling. Likewise, new acquisition techniques arise, driven by new processing algorithms.

7.5.1 Temporal Bandwidth

The importance of this component can be demonstrated with in Figure 7.9, which compares wavelet resolutions on two reflectors 200 milliseconds apart in time. The

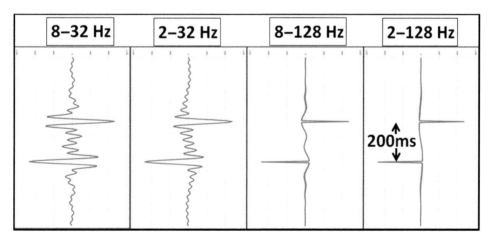

Figure 7.9 Temporal bandwidth and resolution (2, 4, 4 and 6 octaves).

effects of the lack of low frequencies and the lack of high frequencies are both illustrated here.

7.5.2 Noise Reduction

The second major component in achieving 'broadband' marine seismic involves techniques which reduce noise, which can be of several types. For example, ambient noise includes that which is inherent in towing detector cables through the water, plus any kind of 'industrial' ambient noise which could be acoustic interference from other seismic vessels, drilling activity, pipe-laying and so forth.

In addition, there is what could be called 'geological noise' which consists of seismic data originating from the seismic source but which cannot easily be 'imaged'. This can consist of multiple reflections arising in the water layer and in deeper strata, refracted arrivals, reflected refractions, mode converted data and scatter from near-surface irregularities. These 'noise' events although mostly difficult (often impossible) to image and invert, can occasionally be used, and can then be referred to as 'signal'. They are dealt with by a combination of improved spatial sampling of the wavefield including additional sampling in the azimuth domain, by combining hydrophone (pressure) and motion detectors, and by advanced processing algorithms many of which have been developed thanks to improved spatial sampling in the data acquisition. The inversion of geological noise into useful signal remains an ongoing and active research activity.

Populating the azimuth domain is common on most land 3D seismic surveys by virtue of typical survey geometry, but a conventional marine seismic survey is essentially narrow azimuth (NAZ). Azimuth sampling on marine surveys was slow to be implemented due to the high costs involved either in the use of additional source vessels, or by recording additional sail line azimuths, or by using a circular 'coil' towing geometry. However, it is now commonplace (although expensive) and solutions are provided in the text that follows.

7.5.3 'Broadband' Modifications to Conventional Towed Streamer Work

With towed streamers, some of the limitations listed earlier can be overcome, allowing improvements in data quality to be obtained. For completeness these are described here and have trade names or common usage names such as Broadseis^{TM}, Stagseis^{TM}, GeoStreamer^{TM}, GeoStreamer X, GeoSource^{TM}, Orion ^{TM}, Coil^{TM} and FreeCable^{TM}.

The Broadseis towed streamer technique uses a variable depth of tow of the cable(s). In this way, the spectral notches occur at different frequencies at the different source-to-receiver offsets, so that when these different offsets are merged

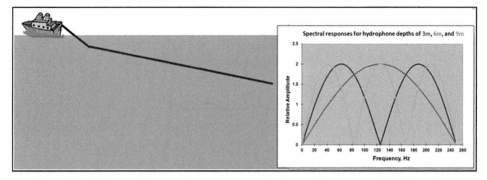

Figure 7.10 Slant cable 3 m deep at near- and 9 m at far-offset end of the cable. The resulting spectral response will be (approximately) the smoothed average of the three responses shown.

(a)   (b)

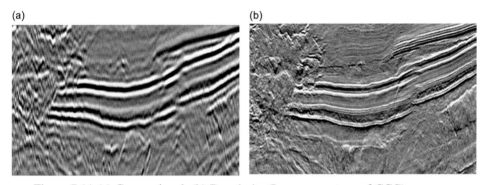

Figure 7.11 (a) Conventional. (b) Broadseis. (Images courtesy of CGG)

in subsequent processing, spectral contributions are received over a wider temporal bandwidth. This was first used by Fairfield in the 1970s and was later named Broadseis^{TM}, marketed by CGG and Shearwater. Figure 7.10 gives a schematic illustration of the technique, which also requires specialist data processing.

Several good examples of improved image resolution resulting from the use of this method have been published, one of which is shown in Figure 7.11.

Another successful technique to fill the detector bandwidth gaps involves adding particle motion sensors to the detector package in addition to the normal hydrophone acoustic sensors. This is illustrated in Figure 7.12.

The downgoing inverted free surface reflection has the same polarity on a particle motion sensor as the original upcoming wave. So with some re-datuming and detector phase compensations, data from the two types of sensor can be merged, such that the upcoming pressure wavefield is reinforced and the downgoing one is attenuated. The simplest way to visualise this is to

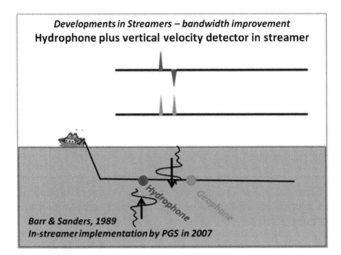

Figure 7.12 Dual-sensor recording.

imagine the two 'traces' at the top of the diagram being summed together algebraically.

This technique is applicable both to towed streamer surveys and to ocean bottom detector systems. In the case of towed streamers, it is the basis of the GeoStreamer[TM] method used by PGS and is also one of several features of the towed streamer IsoMetrix[TM] system developed by WesternGeco and now operated by Shearwater.

A third technique involves the seismic source. Since a fixed source depth imparts spectral notches to the recorded seismic data in the same way as illustrated in Figure 7.8, staggered depth sources can be used. Fired in sequence and 'blended', some spectral recovery is possible. In the case of towed streamer work this is the basis of the GeoSource[TM] technique implemented by PGS. Again, this requires specialised data processing.

A data example using the PGS techniques is shown in Figure 7.13, which illustrates clearly the bandwidth advantages of combining the pressure and the velocity detectors.

7.5.4 Populating the Azimuth Domain: MAZ, WAZ, FAZ, RAZ, AAZ, CLA, FreeCable and WATS

Flexible azimuth sampling is of course inherent in ocean bottom recording, but is also possible with towed streamer acquisition, and the advantages can be very significant.

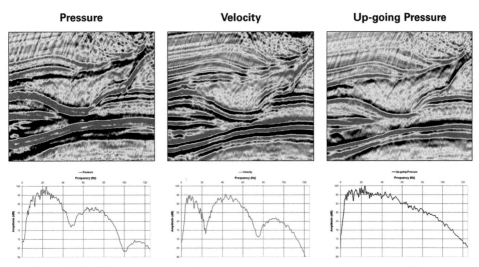

Figure 7.13 Illustrating resolution improvements using dual sensor techniques. (Courtesy of PGS)

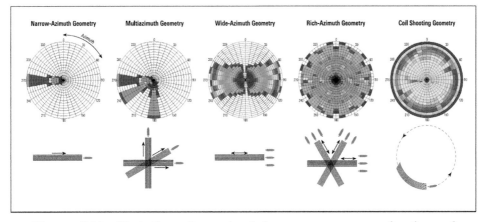

Figure 7.14 Illustrating the azimuthal coverage as a function of acquisition geometry. (Courtesy of WesternGeco)

A conventional towed streamer 3D survey is essentially nearly single or narrow azimuth 'NAZ', as in the left diagram of Figure 7.14. Several techniques can be used to increase the range of azimuths which are acquired with towed streamer, illustrated schematically in the same figure.

First, azimuthal richness can be increased by recording the survey in additional directions.

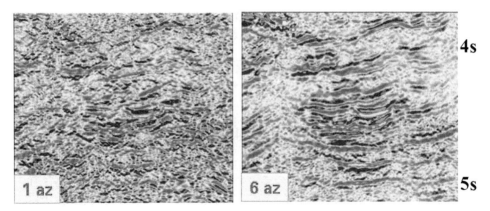

Figure 7.15 Single azimuth versus multi-azimuth. (Source: Keggin, 2007, by kind permission of EAGE)

A striking example is shown here (Figure 7.15) of a multi-azimuth 'MAZ' survey recorded offshore of the Nile Delta by BP. In this case the entire survey was simply repeated at multiple azimuths using towed cables.

In this area one of the main 'noise' sources consists of reflected refractions which can be attenuated by combining a range of source–receiver azimuths. The result was a revelation and triggered a huge interest in populating the azimuth domain on marine surveys.

A schematic example of a very high resolution MAZ technique reported by PGS called Geostreamer X is shown in Figure 7.16. This consists of a single vessel towing a large number of streamers (e.g., 16×7 km length, plus 3×10 km length) with close separations. This combined with three sources, short offsets and three different azimuths then allows a very dense sampling in azimuth, offset and bin size. The long streamers give long-offset data for FWI purposes. The actual geometry is flexible, and the cost is likely to be competitive with OBN.

Next is the technique known as wide-azimuth towed streamer (WATS or WAZ; Figure 7.17). Most commonly there is a main multi-streamer vessel with additional source vessels, and several geometries are possible from the very simple one illustrated here (which might also be used to collect a 3D survey more quickly than using a single vessel). Depending on the actual configuration, WATS is cost competitive with MAZ but with the additional source vessels it is of course more expensive than traditional single- or narrow-azimuth 3D (NAZ) surveys.

There are many geometrical possibilities when using additional vessels. For example, StagSeis™ offered by CGG uses two multi-streamer vessels and three source vessels in a staggered configuration. This is designed to maximise the azimuthal variation, to maintain the offset distribution, and to provide the ultra-long offsets.

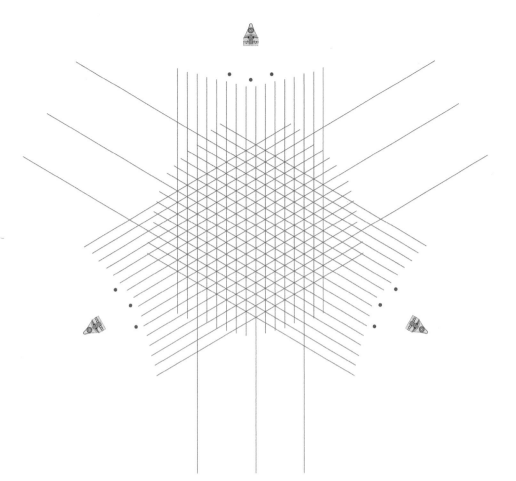

Figure 7.16 Geostreamer X. (Courtesy of PGS)

Figure 7.17 Wide-azimuth towed streamer. (Courtesy of PGS)

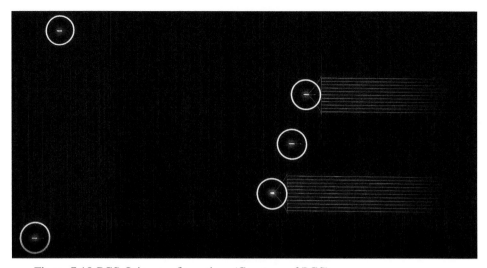

Figure 7.18 PGS Orion configuration. (Courtesy of PGS)

PGS offer the Triton ORION™ configuration which uses five vessels, two of them towing 10 streamers each 10 km long, plus three source vessels (Figure 7.18). These are placed to give source to receiver distances out to over 16 km. When the survey is recorded like this in three different directions, this is MAZ plus WAZ, which gives a full-azimuth result.

Next, Coil shooting (Figure 7.19) is a towed streamer technique first tried in the 1980s (Cole and French, 1984) which has seen a resurgence thanks to improved streamer positioning control and monitoring. The streamer vessels (one or more) progress in overlapping circles, possibly augmented with additional source vessels, giving both a wide range of source–receiver distances and a wide range of azimuths. This is also sometimes also referred to as rich-azimuth shooting (RAZ), and could be full-azimuth (FAZ) with the additional vessels.

A comparison of conventional narrow-azimuth towed cable with coil shooting from the Heidrun field in Norway is shown in Figure 7.20. It has benefited both from dual-sensor recording and rich azimuth sampling (Houbiers, 2011).

Techniques like coil have been made possible thanks to greatly improved steering control for the streamers. This has also enabled vessels to shoot continuously including during their turning from one line direction to its opposite. Using a fairly wide turning circle, this is known as continuous line acquisition (CLA), and both this and coil are advertised by Shearwater.

Unit costs for towed streamer surveys which employ multi-azimuth techniques as described earlier can be expected to be in the $30–$80k/km^2 range.

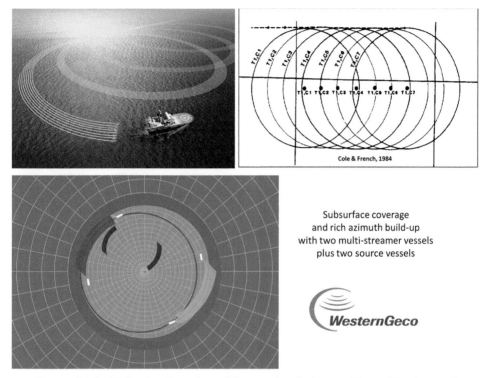

Figure 7.19 Coil shooting to achieve azimuthal sampling. (Courtesy of WesternGeco)

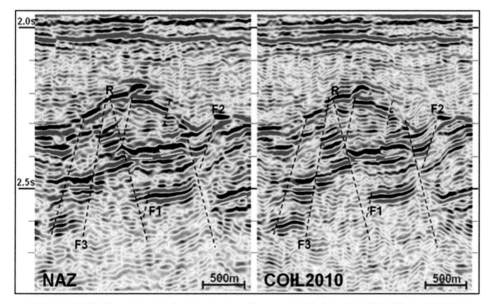

Figure 7.20 Comparison of narrow-azimuth versus coil shooting. (Houbiers, 2011)

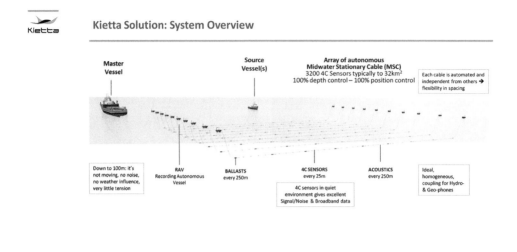

Kietta Solution: System Overview

Figure 7.21 The FreeCable™. (Courtesy of Kietta)

Figure 7.22 FreeCable™. Two of the autonomous vehicles, seen here on a trial in the Red Sea. (Courtesy of Kietta)

A recent development in broadband streamer technology called FreeCable™ (Figure 7.21) has been developed by Kietta, a French company. In this system, a cable with four-component detectors is tensioned at some controlled depth by a pair of marine drones (autonomous recording vehicles).

Up to 10 such cables are envisaged, moving slowly in a predetermined direction, with source vessel(s) as required, all controlled from a master vessel.

This system has most of the advantages of four-component sensors, plus low cable noise (slow moving), azimuth sampling and controlled-depth cables. The drones are substantial, as shown in Figure 7.22.

7.6 Ocean Bottom Detector Systems and Their Technology

Ocean Bottom Detectors and the Reasons for Using Them

Many examples have been published comparing conventional towed cable data with ocean bottom data, and these strongly support the latter. Two early ones are shown here; the first (Figure 7.23) is from the West of Shetlands area. In this case the main benefits arose from the longer source-to-receiver offsets which were acquired with the ocean bottom data, and from the dual-sensor summation. These allowed improved steep dip imaging and better multiple suppression in the target zone.

The second example is from Trinidad (Figure 7.24). Here, the earlier towed streamer data (in a six-cable 5,200 m configuration) had failed to produce satisfactory imaging below the fault. Just two ocean bottom receiver lines 15 km long and 400 m apart produced a substantially clearer image with impressive fault delineation.

Boëlle (2005) compared the signal and noise spectra from towed streamer and ocean bottom data sets from an area in the UKCS. These spectra are shown here (Figure 7.25) in a 1,300 ms window at the reservoir level. By quantifying the area between the red and purple colours as the signal-to-noise bandwidth of the streamer data and between the two blue colours as that of the ocean bottom data, it

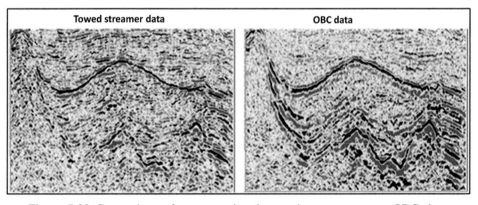

Figure 7.23 Comparison of narrow azimuth towed streamer versus OBC data. (From Kommedal, 2005, with kind permission of EAGE)

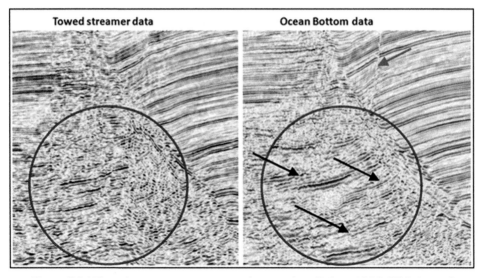

Figure 7.24 Towed streamer versus ocean bottom data (Johns, 2006, SEG Annual Meeting)

can be seen that the separation between signal and noise of the ocean bottom data is significantly better in amplitude and bandwidth.

Deploying seismic detectors on the seabed potentially provides a good technical solution to all eight of the limitations described in Section 7.4 and meets the requirements of 'broadband seismic' for the following reasons:

- The geometry limitations inherent in towing equipment are removed. Thus, a much greater range of source-to-detector distances can readily be acquired.
- Any or all azimuths can be recorded. This allows a step-function increase in the aperture of data acquisition which results in significantly improved noise attenuation, illumination and thus image quality.
- There is no ambient tow noise as in towing streamers at conventional speeds. Figure 7.25 provides a quantification of this.
- The bandwidth advantages of dual-sensor recording are realised, both at the high and low ends of the spectrum.
- Discrimination between upgoing and downgoing wavefields results in attenuation of receiver-side multiples, giving significant imaging improvements over acoustic sensors.
- The source is now operated by dedicated vessel(s) without towed streamers. Source vessels are highly manoeverable and need little time to change direction.
- By using closely spaced source positions, the spatial sampling of the wavefield can be improved.

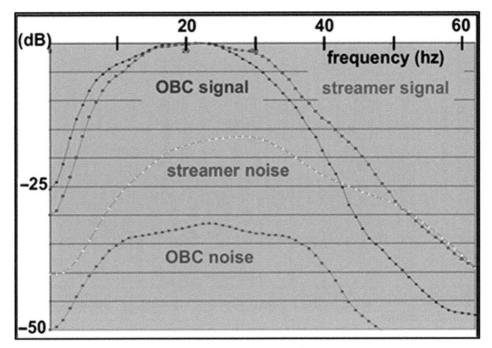

Figure 7.25 Streamer signal and noise spectra compared to OBC. (Boëlle, 2005, SEG Annual Meeting)

- Where platforms or other production facilities limit the area which can be surveyed by towed cable surveys, ocean bottom technology can usually be deployed, providing good quality data in these zones.
- The possibility exists to record and utilise shear-wave data.

There are of course some downsides compared to conventional towed cable work:

- Cost
- Occasional trapped energy noise (Scholte waves)
- Possible coupling issues with the particle motion detectors
- Velocity changes in the water column
- Timing drift with node systems, although this can be corrected

Bandwidth of Ocean Bottom Data

Figure 7.12 gave a simple pictorial description of the main advantage of dual-sensor detector systems – it showed how the inverted downgoing wavefield can be cancelled thanks to the combination of an omni-directional pressure sensing hydrophone with a directional motion sensor.

In the frequency domain, the result is to allow a flattening of the signal spectra, because the spectral responses of the two sensors are complementary as shown in

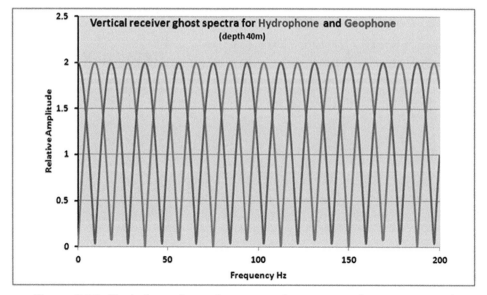

Figure 7.26 Vertical receiver ghost spectral responses for pressure and motion sensors.

Figure 7.26. As the depth increases, the spectral peaks and troughs become closer together, giving effectively a very flat spectrum.

The hydrophone and motion sensor outputs are processed after phase alignment by scaling one, and adding it to the other, taking into account the impedances of the water and the seabed. A simple and robust method of doing this is proposed in a CREWES report (Hoffe, 1999) in which a time-variant smoothed scalar trace is computed for each hydrophone–geophone trace pair before being applied. Plotting the scalars can give an indication of the quality of the coupling of the node to the seabed.

In practice, most recording instrumentation will impose a low frequency cut-off around 1 Hz. Additionally, if the motion sensor is a velocity geophone, this will have a response which falls at 12 dB/octave below its natural frequency (usually between 8 and 15 Hz). This is sometimes pointed out as a detrimental feature of velocity sensors, although in reality the sensor and instrumentation dynamic ranges and noise levels allow recovery several octaves lower if these exist in the data. Microelectromechanical System (MEMS) sensors with extended low-frequency responses are beginning to appear in OBN systems.

OBS, OBC or OBN? Ocean Bottom Seismometers, Cables and Nodes

Ocean bottom recording is not new – earthquake monitoring systems were in use in the 1930s, and some land seismic equipment has been waterproofed and used on the seafloor from the 1950s.

7.6.1 Ocean Bottom Seismometers

The term OBS is usually given to the ocean bottom seismometer systems used mostly by academia and research establishments for microseismic and earthquake monitoring, and for long-offset refraction work for assessment of deep strata, basement detection, and velocity tomography. Details can be found in Chapter 3 of this book.

7.6.2 Ocean Bottom Cable Systems

OBC systems (Figure 7.27) use detectors built into a cable configuration which can be laid out and picked up from a mother ship. OBC systems have been in use in the seismic industry for many years and some deployment continues. However, cable-laying is slow and expensive – so the unit cost of the seismic data is typically around seven to eight times the cost of a conventional towed cable survey.

The deeper the water, the more difficult the operation, mainly due to ocean currents. Cables suffer wear-and-tear during deployment and retrieval and coupling between the detectors and seabed can be poor. Where the seabed is rugose, there is the possibility of a sensor unit suspended without contact with the seabed. Currents can also cause high levels of noise on the motion detectors due to cable strumming, and as with OBN systems, some Scholte wave noise may be experienced. Earlier OBC geophone units used gimbal mountings to ensure a true vertical direction of the z geophone. Gimbals themselves have a transfer function and may therefore impart noise or distortion to the seismic signals, and it became

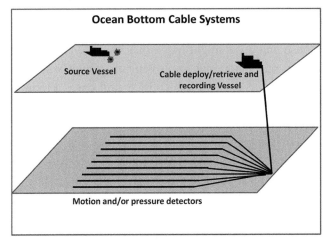

Figure 7.27 Ocean bottom cable systems.

preferable to avoid gimbals and to orient the geophones using first arrivals or to use MEMS accelerometer sensors which can sense the vertical direction.

Data from the seabed detectors (which can be hydrophone plus vertical motion, or hydrophone plus three-component motion) are transmitted through the cables and recorded on the mother ship.

In volume terms OBC work has now been overtaken by OBN systems, although see later under permanent reservoir monitoring PRM systems.

7.6.3 Ocean Bottom Nodes

OBNs and OBSs are essentially the same things – both are autonomous recording systems which are deployed onto the seabed, record data, and are eventually brought to the surface and the data recovered. Current usage favours the terminology OBN for reservoir seismic purposes.

Since around 2000, lower power electronics, higher capacity batteries, improved deployment and recovery techniques and fast-expanding electronic data storage capabilities have brought major changes in design. These have allowed OBN systems to become popular for 3D marine seismic surveys, in which the nodes have to stay in position and stay 'live' for around a month or more.

The successful examples of ocean bottom data such as those shown in Figures 7.23 and 7.24 have been sufficiently compelling that the OBN market is growing fast despite the relatively high cost of surveys. The unit cost of 3D seismic surveys using OBN systems is currently in the range $35k–$200k per km^2, with the lower part of this range applying to shallow or intermediate water depths down to around 400 m where the use of highly automated systems is possible. Costs, however, are dropping and are likely to come down further as improved deployment techniques are developed, and 'blended sources' become routine. One estimate is that these unit costs might halve over the 10 years from 2015 to 2025. The current focus is on operational improvements which will both increase the spatial sampling of the surveys and reduce their unit costs. Production rates over 20 km^2 per day have been claimed, which is comparable with a large multi-streamer towed system.

By around 2010, companies offering OBN services were advertising the availability of inventories of typically 1,000 units. A few years later, that number had multiplied 10-fold, and crews are currently mobilising with considerably in excess of 10,000 nodes. Companies are advertising with slogans such as 'Dense Seismic on the Seafloor'. The OBN market has actually been stimulated by a trend among some of the major seismic contractors to become 'asset light', in other words to avoid owning and operating expensive hardware such as seismic vessels and nodes. This has encouraged new specialist node companies, a node rental market and the development of handling systems which are being designed to be fitted relatively easily onto vessels of convenience, and to handle more than one type of node. The phrase 'node agnostic' describes some of this market. This flexibility is beginning to allow the emergence of multi-client OBN work.

Node Design and Deployment

Deployment techniques depend heavily on the water depth and on the environment of the survey area.

In very shallow water or transition zones, nodes can be deployed from a very shallow draught Z-boat or similar (Figure 7.28).

The nodes themselves could be attached to a rope system for retrieval, or they could be individually buoyed, or use a pop-up design.

As the water depth increases to several metres or more, a more conventional vessel of convenience would be used, with many nodes on a flexible rope system. The nodes here in Figure 7.29 are by Geospace, which manufactures and rents them.

In deeper water, the deployment methods depend on factors such as tides and currents, and on the presence or absence of production infrastructure. In open waters, vessels such as the one in Figure 7.30 can lay and retrieve nodes using an automated conveyor-belt system. The nodes stay attached to a steel cable which dictates the node spacing on the seabed and allows fast retrieval.

Once deployed, the three directional components of particle motion are calibrated using a combination of inbuilt tilt sensors and compass, plus the use of direct arrivals from the many known source points at many azimuths and offsets.

Figure 7.28 Node deployment in very shallow water. (Courtesy of Geonode, Latvia)

Figure 7.29 Deployment of third-party nodes in shallow water. (Courtesy of Empress Marine, Dunsborough, WA)

Figure 7.30 (a) Nodes-on-ropes deployment from a specialised vessel. (Courtesy of Magseis Fairfield). (b) Node deployment from a specialised vessel. (Courtesy of Magseis Fairfield)

This 'nodes on ropes' technology is now the most common for intermediate water depths, down to around 700 m, and it has made really significant progress during the past 10 years. Typically, these are fully containerised solutions which can be installed on vessels of convenience. They combine deployment, retrieval, data extraction and management and battery charging systems. This one (Figure 7.31) by InApril, for example, claims to be able to deploy up to 10,000 nodes from a single vessel, with deployment and retrieval speeds of up to 6 knots.

Actual average deployment and retrieval speeds reported by the industry for such systems are of course considerably less, but it is technologies such as these which are generally increasing survey speeds and reducing costs.

Node designs and construction depend on the maximum water depth to which they are specified to operate. Most OBN manufacturers offer two or more models for different water depth ranges. The shallower versions have maximum depth

Figure 7.31 Node deployment, containerised solution. (Courtesy of InApril)

Figure 7.32 Range of nodes from InApril. (Courtesy of InApril)

specifications in the range 10–700 m, with the deeper versions allowing operation down to around 3,000 m.

The Venator™ range by InApril shown in Figure 7.32 is designed for both rope or ROV deployment, and has models for 400 m and 3,000 m depths, with battery life options for 50, 110 and up to 200 days.

The schematic in Figure 7.33 illustrates a typical deployment layout in medium depths of water. On completion of the shooting, the acoustic release of a buoy allows the mother ship to commence retrieval. A large operation of this type will use two node handling vessels plus three source vessels, and one or two support vessels.

Figure 7.33 Node geometry example. (Courtesy of Magseis Fairfield)

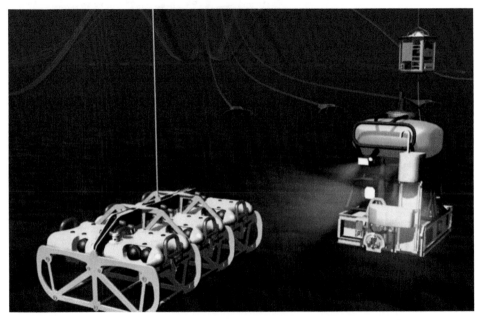

Figure 7.34 Node deployment by ROV. (Courtesy of Seabed Geosolutions)

As the water depth increases, or in the presence of infrastructure, or where the currents are strong, deployment and retrieval of OBN systems usually takes place using remotely operated vehicles (ROVs; Figure 7.34). This increases the cost considerably, as an ROV support vessel typically costs around $180k–$250k/day, and operating speeds are slow relative to those of 'nodes on ropes'. However, an ROV vessel is capable of deploying multiple ROVs simultaneously, which reduces the effective deployment cost. Deployment rates of 600–1,250 nodes per day are possible.

ROVs are deployed with baseline acoustic positioning systems which can locate the nodes within close tolerances. The ROV also 'plants' the node to ensure good coupling with the seabed, as in this shot of the Manta node (Figure 7.35). This node, suitable for ROV or 'on-rope' deployment down to 3,000 m, can also be deployed in multi-ROV configurations from a single support vessel.

Magseis Fairfield also offers a wide range of nodes (Figure 7.36).

Figure 7.35 Manta node. (Courtesy of Seabed Geosolutions)

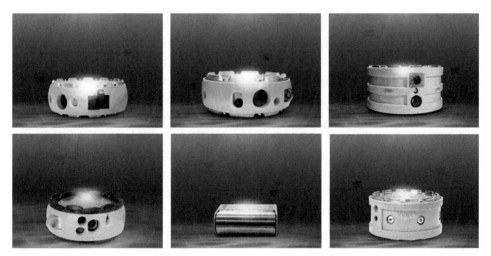

Figure 7.36 (Top row) Suitable for 100 m, 700 m and 3,000 m. (Bottom row, left to right) A multipurpose node capable of dual methods of deployment (down to 4,000 m), their Marine Autonomous Seismic System (MASS) node-on-rope (0–3,000 m), and on the right is the ZLoF™. This offers a 5-year deployment life, with a battery which is adequate for 500 cumulative days of recording within the 5-year period. Data can be extracted optically using an ROV. The intention is to provide a competitive solution to a PRM 'Life of Field System' for time-lapse recording. (Courtesy of Magseis Fairfield)

Figure 7.37 GPR ocean bottom node (Courtesy of Sercel)

Figure 7.37 shows a new entry (September 2019) by Sercel/CGG in collaboration with BGP. This utilises Sercel's GPR Quietseis[TM] detector package which has MEMS sensors rather than the more common conventional moving coil geophone.

Repeatability of Nodes for 4D Surveys

Testing with synthetic data suggested that time-lapse surveys could be successfully undertaken using nodes, although care should be taken with several issues including the effects of water column velocity variation. Subsequently, several node-on-node 4D surveys have been successfully acquired. One of the earliest and well-documented deep water OBN surveys was conducted by BP on the Atlantis field in the Gulf of Mexico and is described later in Section 7.10. This area included the Sigsbee escarpment, where water depths varied from around 1,300 m down to 2,200 m. Several repeat OBN (time lapse) surveys have now been carried out over this field with successful outcomes (Reasnor, 2010), and other such surveys have been described (Stopin, 2011; Boëlle, 2012). To quote from Stopin, 'The quality of the data and the processing applied led to a high repeatability with a normalised root mean square (NRMS) of around 6%. Proper correction of the shot statics related to water temperature variations is essential if depth shifts are used to infer geo-mechanical effects.' Six percent NRMS is a satisfactory number.

Highly accurate seabed positioning using ROVs can now be provided by acoustic transponder baseline systems and by data analysis of first arrivals. For rope and other deployments, careful analysis of direct arrival data is used and is generally satisfactory.

AUVs: Robotics, Flying Nodes and 'Swarms'

There has been considerable interest in this technology for several years, both in academia and by vendors. A flavour and some actual examples of the technology are shown in Figures 7.38–7.42. It has been driven by the growth in complexity of the various marine techniques required to acquire 3D, 4D and especially azimuthal seismic data. To quote from the Delphi project at Delft University, 'From … acquisition trends it is clear that the current approach for seismic surveying will become a logistical nightmare. Therefore, robotising seismic data acquisition seems an inevitable next step.'

There are other major attractions, for example, to place AUV nodes very close to seabed infrastructure.

One of the first commercial prototypes to appear was the SpiceRack^{TM} initiated in 2012 by CGG (then CGGVeritas) in collaboration with Saudi Aramco (Figure 7.38).

SpiceRack^{TM} development has continued with Seabed Geosolutions, and recent trade shows have reported a successful test of 20 nodes in the Red Sea in 2017. This device is being engineered for depths down to 3,000 m.

There are also some EU-funded projects which address the development of 'swarms' of autonomous nodes.

The one shown in Figure 7.39 is entitled 'Large-scale piloting and market maturation of a disruptive technology comprising a fully automatic survey system

Figure 7.38 'SpiceRack' AUV node. (Courtesy of Seabed Geosolutions)

Figure 7.39 EU-funded Project Cordis (https://cordis.europa.eu/project/rcn/204237/reporting/en).

Figure 7.40 The EU 'WeMust' AUV development (http://wimust.isme.unige.it/).

dramatically reducing the operational cost of handling swarms of autonomous sensor nodes.'

Another EU-funded project is WiMUST (Widely Scalable Mobile Underwater Sonar Technology), which has explored AUV robotics and demonstrated some shallow water capability (Figure 7.40).

To quote from their website, 'The robots swarm together and explore using sonar technology. One of our objectives was to have the greatest possible number of these vehicles underwater. Every robot is equipped with an atomic clock, something that some years ago was basically impossible. This is what allows our robots to navigate underwater. The next challenge is how they control their navigation in coordination with the others. They know where they are thanks to geo-location, so the issue here is how they work in relation to each other' says Luis Sebastiao, a sound engineer from the University of Lisbon.

The company GoScience, now called Autonomous Robotics Ltd, was an early implementer of flying nodes for seismic surveys.

The AUVs 'fly' to a predetermined position, deposit a multi-component detector, record data and eventually return to a mother ship with the data.

A successful test has been reported offshore Plymouth in October 2018, and the company is collaborating with Robert Gordon University Aberdeen in the development of 'swarm' technology.

A look at recent patents suggests that the AUV technology could be developing strongly over the next few years. One project, called Flatfish, is a joint venture with Shell. It is likely to be a multi-purpose AUV with inspection and possibly seismic duties. For example, it might extract data from seismic nodes and charge their batteries. There is considerable interest in increasing the autonomy of ocean bottom nodes, enabling them to be deployed over large distances on the seafloor and remain over long periods. They could be ready to record seismic and other data on demand, for example, for 4D surveys.

Another recent reference is 'Design, modelling and imaging of marine seismic swarm surveys' (Muyzert, 2018, with later corrections).

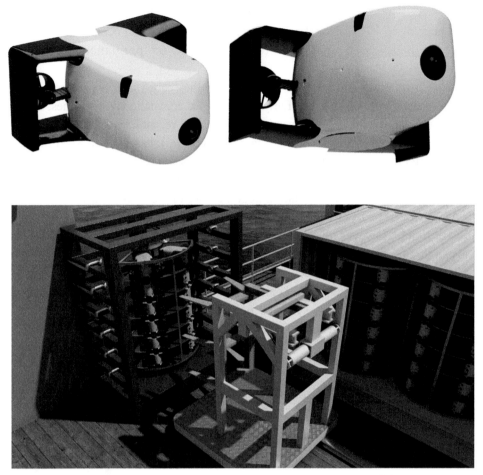

Figure 7.41 AUVs and their robotic handling systems. (Courtesy of Autonomous Robotics Ltd)

Figure 7.42 Already well developed for airborne drones, the same is likely to happen with AUVs. The future? Possible 'swarms' of intelligent nodes.

Permanent Reservoir Monitoring Systems

Commencing in 1995, several grids of permanently emplaced ocean bottom cable systems have been installed for long-term or permanent reservoir monitoring (PRM).

These consist of OBC systems trenched into the seabed and can utilise several types of sensor and transmission. Early systems used conventional geophones, digital conversion and electrical data transmission. The more recent installations tend to use optical fibre for sensors and transmission, as at Ekofisk, Jubarte, Johan Sverdrup and Johan Castberg reservoirs. The high up-front cost has so far limited major expansion of these systems, so they have been deployed on reservoirs which will require a large number of monitor 4D surveys and a large number of wells to be drilled.

As can be seen from Figure 7.43, it is not yet clear whether this is a strong growth trend. However, the data published so far on these systems demonstrate very high data quality and the best possible repeatability for '4D' surveys.

The company Geospace Technologies has been the main supplier of electrical systems. An optical system called called OptoSeis^{TM} was developed by PGS, used in the Jubarte and Ekofisk systems, now purchased by Geospace. An optical system developed out of a QinetiQ company and called Stingray^{TM} did not survive the 2014 downturn.

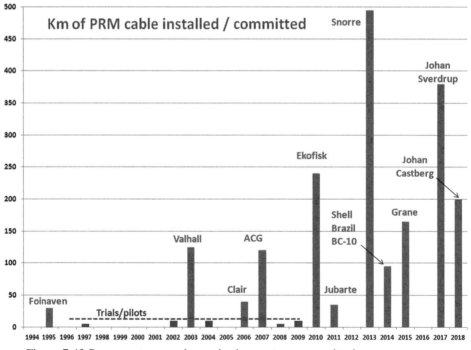

Figure 7.43 Permanent reservoir monitoring systems committed.

Alcatel, a well-known company in trans-ocean cabling, but a relatively new entry in the seismic market, has supplied two of the most recent PRM optical fibre installations, as depicted in Figure 7.44.

Another supplier offering PRM systems is Octio in Norway.

A variation on PRM systems is SoundSabre^{TM} (Figures 7.45 and 7.46). This is not yet commercially available as of 2020 but is described here for completeness.

SoundSabre is a node-based permanent monitoring system for offshore fields, to acquire 4D seismic monitor surveys and passive microseismic data.

Figure 7.44 A view of the Alcatel optical fibre installation on Johan Sverdrup field. (Courtesy of Equinor)

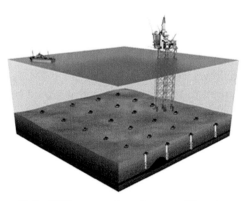

Figure 7.45 SoundSabreTM, general scheme. (Courtesy of Giles Watts, SoundSabre)

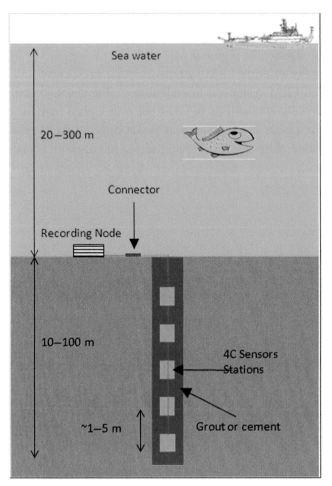

Figure 7.46 SoundSabre™, vertical detail. (Courtesy of Giles Watts)

It consists of a number of vertical arrays of detectors installed in the seabed above a reservoir.

It might offer much lower costs than alternative PRM systems (no trenching, and no connection required to the platform), and it would be more sensitive than seabed systems thanks to its subsurface installation. It would contain low-frequency detectors to capture natural seismicity.

There have been several recent trials of distributed acoustic sensing (DAS) systems which use Rayleigh scattering of light signals within a fibre as an indication of movement such as a seismic disturbance along the fibre (Figure 7.47; see, e.g., Farhadiroushan, 2019). A number of installations exists both for downhole vertical seismic profiles (VSP) and for surface seismic, and these of course fall into the category of PRM systems.

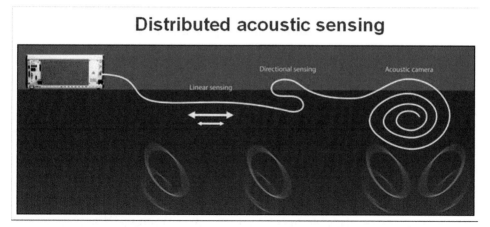

Distributed acoustic sensing

Figure 7.47 Distributed acoustic sensing via fibre-optic. (Parker, 2014, By kind permission of EAGE)

A logical step is the integration of the downhole data with the seabed data in PRM systems.

7.7 'Blended' Sources and Full-Waveform Inversion

Blended Sources

Oil companies had noted the very significant cost and productivity improvements in land seismic which resulted from a combination of 'blended' source techniques and newer detector systems such as land nodes and had been pushing for similar experiences in the marine environment.

By 'blended sources' or 'simultaneous sources' is meant the activation of the seismic source(s) during the normal recording time window needed by a previous source activation. Some kind of randomisation is normally used, either in shot timing such as 'dither', or in position, or both. In general, the greater the degree of randomness of the interfering' sources in the acquisition design, the greater the success in de-blending the resulting data (Hampson, 2008; Abma et al., 2015).

The key to the technology is the ability to separate or 'de-blend' the two (or more) interfering shots recorded by the receivers. Several data processing methodologies are possible, one of which involves a domain translation from common shot to common receiver, thus rendering the 'interfering' shots into random incoherent events in the new domain, which can then be suppressed. Sparse inversion is developing strongly as the newer processing methodology of choice for de-blending the data.

In this way blended sources are increasingly being used to record cheaper and higher quality offshore surveys, utilising additional source vessels which are relatively cheap and much more manoeverable. The receivers can be nodes or a multi-streamer vessel configuration. For examples of configurations for FYI work, see Roende (2019) and Brenders (2018).

Full-Waveform Inversion

A trend exists towards improved velocity modelling and subsequent data imaging in complex tectonics using FWI. To be successful this utilises very long source-to-detector distances (e.g., up to 40 km), and an improved source output at very low frequencies down to around 1.6 Hz. Nodes lend themselves well to this trend, especially those with good battery longevity, see for example Roende (2019) and papers to be presented at the EAGE 2020. The development and use of a very low frequency source specially designed for FWI called Wolfspar^{TM} is described in Brenders (2018), and by Dellinger (2016, 2019).

7.8 OBN for Shear-Wave Seismology

One of the benefits of multicomponent ocean bottom technology is its ability to record the shear wave arrivals on the horizontal components on the motion sensors. These do not propagate through the water layer.

In practice, the data processing of shear wave data is very many times more complex than that for P-wave, so this technology exists in a very small 'niche' market. However, two examples are given here, with their context.

Several reservoirs exhibit leakage of gas called 'gas chimneys' from the reservoir up into the mid-section. This results in scattering of the p-wave data. By utilising converted shear waves (converted at top reservoir), which are relatively unaffected by the gas or fluids in the 'chimney', some interpretation of the reservoir may be possible. Good examples have been shown for both the Tommeliten and Valhall fields in the North Sea, the one in Figure 7.48 from Valhall in Johnston (2010).

In this example from the Alba field, Figure 7.49, the p-wave reflectivity at top reservoir is very small. However, there is a good shear-wave response, hence the results from an OBC survey were spectacular.

There has been discussion about whether a long-offset p-wave survey might provide equally good data from amplitude variation with offset (AVO) information, but this OBC example is impressive nonetheless.

Fractured reservoirs are those which should derive most benefit from shear wave seismic work. This is currently also a niche market but may expand in future. It will be stimulated as OBN or other robotic OB surveys increase their market share, since they all routinely supply shear wave data.

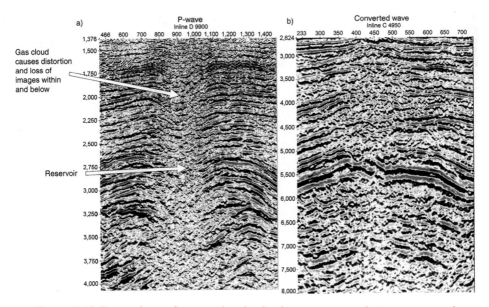

Figure 7.48 Comparison of conventional seismic versus ocean bottom converted wave data, in Johnston (2010). (Courtesy of SEG)

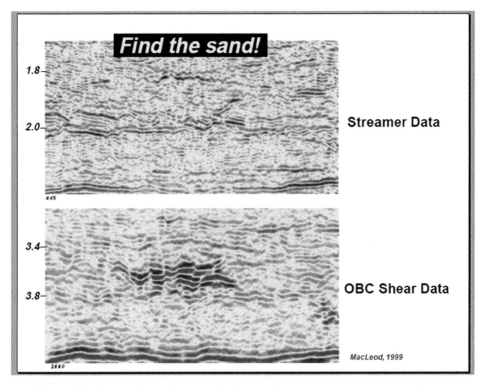

Figure 7.49 Comparison of conventional seismic vs ocean bottom converted wave data. (From MacLeod, 1999, *The Leading Edge*, SEG)

7.9 Value of OBN Technology

Quantifying the value of seismic data has tended to lag extensive use of the seismic method itself. In the 1980s, 2D seismic was the norm, and 3D was regarded as a luxury. Gradually, 3D became an established part of exploration and development, and by around 2000, management would generally refuse to sanction a well if its trajectory had not been derived from 3D seismic data. Increasingly, similar strategy is now applied to reservoir management, such that repeat seismic (4D) has routinely become part of the reservoir depletion plan.

Exploration, however, is moving into more and more difficult areas such as deepwater plays, and these are beginning to see a significant use of 'broadband' seismic methods. Since around 50% of the marine seismic market is now 'broadband', oil companies are clearly prepared to pay the high cost premium to acquire these improvements in data quality.

This perspective of where the seismic process features in the overall reservoir development capex (Figure 7.50) is one enabler for OBN technology, since this arguably produces the highest possible quality image of the subsurface.

One way of assessing value is to quantify it via a decision tree approach, as in Houck (2010). He takes a model-based approach to value of information (VOI) which can be adapted to the changes in value of a data set during field life. The example here (Figure 7.51) is for repeat seismic (4D) but is adaptable to any seismic data set.

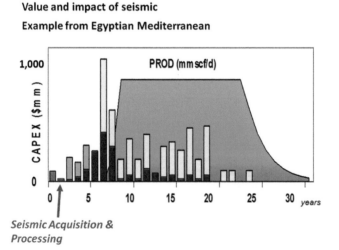

Figure 7.50 Cost and value leverage of seismic work. (Courtesy of J. Pape)

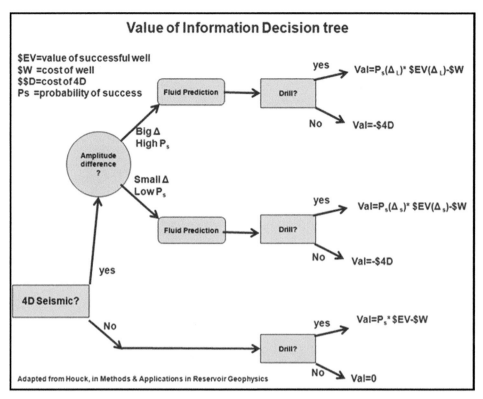

Figure 7.51 Decision tree. (Adapted from Houck, 2010)

7.10 Case History: The Atlantis Field

Summary This deepwater subsalt reservoir has been an important and rewarding laboratory for BP's seismic technology over several years. The $1.3 billion Atlantis Phase 3 development (2019) was approved thanks to the company's seismic imaging and reservoir characterisation breakthroughs which have revealed an additional 400 million barrels of oil in place at the Atlantis field.

Atlantis is a deepwater subsalt reservoir in the Gulf of Mexico in the southeastern Green Canyon area, about 300 km south of New Orleans. It was discovered in 1998, with first oil production in 2007. It lies partly beneath the Sigsbee Escarpment, where the water depth changes from 1,300 m down to 2,200 m. The central business challenge underpinning the Atlantis OBS node project was the poor and unreliable seismic image quality over the majority of the reservoir area. This was due to several issues – surface-related multiple contamination, uncertainty in the sediment and salt velocity model, complex illumination due to

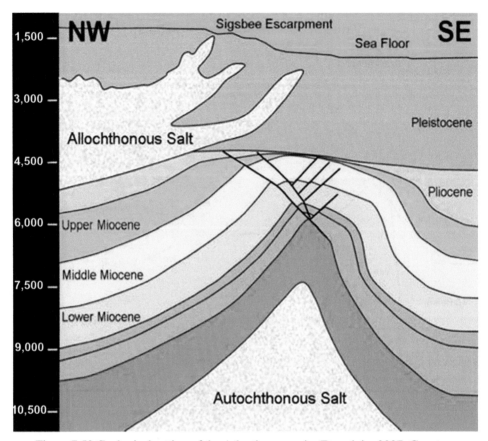

Figure 7.52 Geological setting of the Atlantis reservoir. (Beaudoin, 2007. Courtesy of SEG)

the overlying salt geometry and the steeply dipping water bottom associated with the Sigsbee Escarpment. A complicated salt body with multiple salt fingers covers the northern part of the field, making imaging particularly difficult. Then, on the southeast side of the field, the limb of the anticline has extreme dips (more than 60°) which are also a challenge to image. Only the southern part of the field can be imaged with confidence. The three reservoirs are 17,000 ft below sea level and consist of poorly consolidated Middle Miocene turbidite sands (Figure 7.52).

Up until the mid-2000s, reflection seismic work had been undertaken by streamer with various geometries. By 2005 the WATS technique (one multi-streamer vessel and several shooting vessels), with its rich diversity of azimuth and offset, had been the most effective in improving the imaging quality.

At that time nodes were generally considered to be the option of last resort on account of availability and cost, especially for deep water. However, multi-azimuth acquisition was being successful on land, and was taking place via WATS and

Figure 7.53 View of the seabed area above the reservoir. Each red dot is a node position, above, on and below the Sigsbee Escarpment (Beaudoin, 2007).

WAZ offshore, so in 2005 more than 900 autonomous nodes were commissioned for a major new survey at Atlantis. These were required to be positioned by ROV over a 16 × 16 km grid. The survey design and logistics were constrained by the inventory of nodes, the speed of placement and retrieval, and by the battery life which at that time was limited to about 28 days. These various constraints dictated that the survey be recorded in two ''patches' of nodes, which required duplicating many of the source locations. However, to have avoided this and to have recorded the entire survey in one patch would have required 1,600–1,700 nodes as opposed to 900. In practice the main critical parameters for operational success were node battery life and the operational speeds of ROVs and the source vessel. (Node laying speeds were 0.6 knots maximum with the average being half of that.)

The node geometry was based on modelling using the best knowledge of the salt geometry available at that time, as well as the critical operational constraints mentioned earlier. Node spacing was on a hexagonal grid with 426 m spacing (6.36 nodes/km^2). Positional control was initially with the ROV's inertial and acoustic system, and then subsequently using direct arrivals from the source positions from which were also derived the orientation and vector fidelity of each node. Ninety-five per cent of the nodes were located within 10 m and 99% within 20 m of their desired locations. Shots were fired on a 50-m grid and at an 11.2-second interval (although the nodes recorded continuously). Each node contained three orthogonal geophones (without gimbals), a hydrophone detector and a tilt metre.

A comprehensive description of this survey can be found in Beaudoin (2007).

Early imaging results from this survey used the existing velocity model derived from a narrow azimuth streamer survey but were very promising under the salt canopy, as can be seen in Figure 7.54.

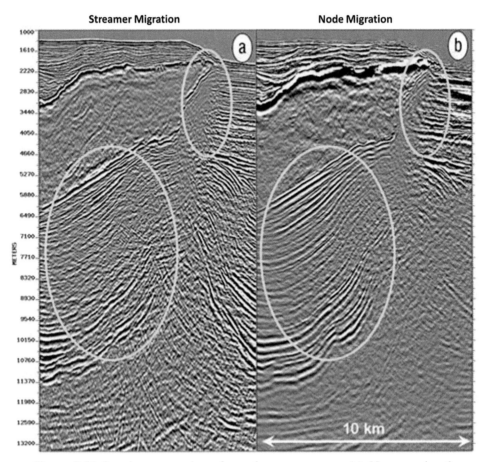

Figure 7.54 Imaging comparison from the survey core. (a) Narrow-azimuth streamer. (b) Node data. The ovals show subsalt regions where nodes are providing improved images of the subsalt north flank and immediately under the leading edge of the salt canopy (Beaudoin, 2007).

The 2005/2006 node survey had demonstrated that nodes could be positioned using ROVs with good accuracy in deep water and in difficult conditions. In 2009, an ROV was used to position 500 nodes to a subset of the original 1,628 node positions. Since first oil had commenced in October 2007, one of the objectives of this new deployment was to test the viability of nodes for time-lapse data. Different nodes were used but they had previously been compared in a separate deepwater test. The survey 'patch' included part of the Sigsbee Escarpment, so the large water depth variation and the seabed rugosity encountered here was expected to be challenging for positional repeatability. Also, the new set of nodes used external sensor coupling, which meant that some latitude in positioning was allowable.

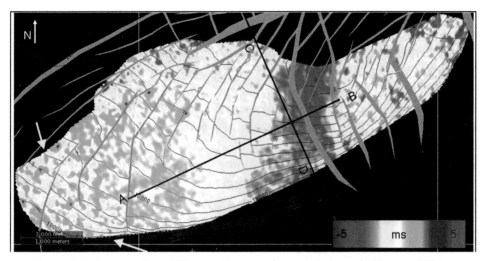

Figure 7.55 Average time shift extraction map from a window of 100 ms to 250 ms above the reservoir section. Brown faults and depth structure of top reservoir are overlain. Black lines show location of a seismic line in the referenced publication (van Gestel, 2013). Yellow arrows indicate the area dominated by remnant multiple.

Despite these issues, 91% were deployed to within 5 m of their baseline positions and 98% within 10 m.

Almost all of the nodes (98.8%) provided good data on return. In the good signal areas, time-lapse repeatability (with an NRMS of 5.3%) was as good as had been experienced on any other of BP's time-lapse surveys, although some additional processing was required due to some bandwidth differences and seabed conditions. Repeatability was affected by differences in multiple period due to water velocity variation, and this was troublesome in areas of low signal. However, NRMS values around 6% were subsequently reported elsewhere on similar surveys (Stopin, 2011), who reported that care was needed with water time variations.

A major conclusion from this first-ever node-on-node 4D survey was that ROV-deployed node technology could be used to reliably conduct highly repeatable 4D surveys. As an example, Figure 7.55 shows the time shift map from a region just above the reservoir section. It has good correlation from well pressure data.

These rich data sets allowed extensive velocity analyses, resulting in the derivation of anisotropic velocity models for the Atlantis field (Sil et al., 2009). The node data also allowed identification of prism wave generation (reflections from one salt body reflected again from another), which were subsequently successfully and usefully imaged and interpreted.

The success of this 4D survey raised the possibility of utilising autonomous nodes for repeat surveys rather than permanently deployed (PRM) systems, and at least it started technical and financial discussions around this subject.

In 2009, a trial of marine blended sources was also recorded into nodes at Atlantis. Nodes, since they record continuously, are a natural fit for blended source technology, which had been working satisfactorily on land for several years. The trial survey lasted for 3 days, using two source vessels shooting simultaneously and independently, incorporating some deliberate randomness. Completing the same survey with conventional acquisition would double the survey time, since only one source vessel would be in operation at any one time. The test was performed in conjunction with a node survey using conventional source operation such that a direct comparison was available (Figure 7.56). The blended source signals were successfully separated by a sparse inversion algorithm, as can be seen from this result in Zhang (2013).

Water injection in the reservoir commenced in 2011, and so there was a business need to monitor its direction and extent after a suitable period. By 2014/15, many more nodes had become available (1,912, which is more than the 1,628 locations used in previous surveys). Also, battery life had improved such that the nodes could be emplaced for longer periods. These 1,912 nodes were deployed and retrieved over an area of 289 km^2 at the Atlantis site over a period of 151 days (including source acquisition). The average node emplacement duration was 85 days. The maximum source–receiver offset was increased to 10 km. These longer offsets were used to improve velocity model building via FWI. This survey was satisfactorily completed, although node deployment was affected by loop currents which can be severe in the Gulf of Mexico (surface currents up to 4 knots have been encountered). As a result, node deployment and retrieval were limiting factors.

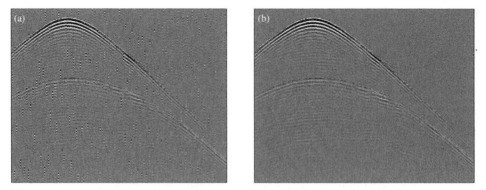

Figure 7.56 (a) Blended, unseparated. (b) Separated (Zhang, 2013).

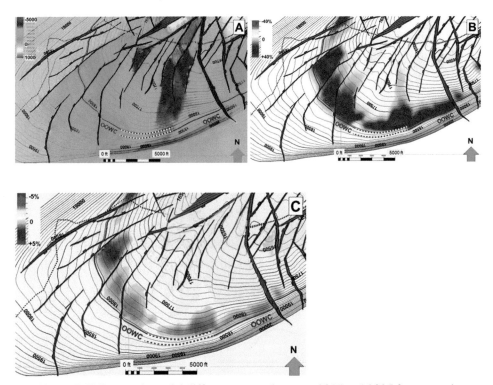

Figure 7.57 Reservoir model difference maps between 2015 and 2005 for reservoir B. Brown outlines indicate faults, contours are in black, dark blue line is OOWC. (A) Pressure difference (psi). (B) Water saturation difference (%) (C) Modelled Extended Elastic Impedance (EEI) difference (%) (VanGestel, 2017).

This second monitor survey allowed confirmation of the movement of the recent water injection program, and also gave good indications of the original oil water contact (OOWC) location. Time-lapse changes were also visible in the difficult region below the salt, Figure 7.57.

The time-lapse surveys have also been analysed for geomechanical effects, locating the regions of stress arching. This allowed more accurate interpretation of the time changes observed at reservoir level.

It had become apparent that extending the low-frequency seismic source bandwidth to overcome ambient microseismic energy below about 2 Hz was challenging using airgun sources. BP therefore designed and built a very low frequency source called Wolfspar which could sweep from 1.6 to 8 Hz and could be detected at offsets up to 30 km. This was tested at Atlantis. It has assisted in imaging through the complex overburdens and is used in velocity model building in the FWI process.

Figure 7.58 The Wolfspar seismic source. (Dellinger, 2016 . SEG Annual meeting)

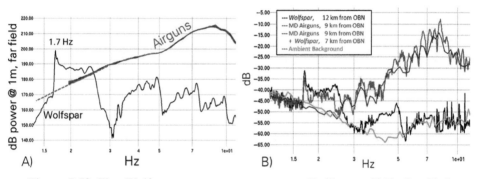

Figure 7.59 The Wolfspar source power spectra. (Dellinger, 2019. By kind permission of EAGE)

A picture of this large appliance is shown here in Figure 7.58, with the power spectral plots in Figure 7.59.

The next development concerns VSPs. Normally, acquiring 3D VSPs with down-hole geophones uses expensive rig time (typically 3–8 days per well). To reduce these high costs, BP has used permanently installed distributed acoustic sensing (DAS) optical fibres in the Atlantis Phase 3 Field Extension. Since these are subsea wells, several kilometres of fibre cabling are required, but without any subsea electronics. In this case the DAS interrogator is located on the semi-submersible platform production quarters (PQs) and connected to the subsea well through a fibre-optic umbilical and series of subsea umbilical termination assemblies (SUTAs) and an Optical Distribution Unit (ODU), similar to the configuration shown in Figure 7.60.

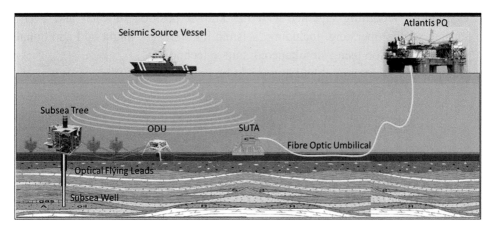

Figure 7.60 VSP data using DAS. (Naldrett, 2020. By kind permission of EAGE)

As can be gathered from this case history, Atlantis has been used over many years as a laboratory for extensive at-scale seismic experiments of many types both in acquisition and data processing, and many important conclusions can be drawn from them. Some of these are listed here, with some possible extrapolation added.

- Node positioning is easier than sensors on streamer surveys.
- Tow noise is non-existent, and the sea floor is a very quiet environment.
- Varying source characteristics are easier to identify and to correct for.
- Deepwater OBN data, however, are strongly susceptible to the changes in speed of sound in water caused by differences in salinity and temperature as loop and eddy currents bring 'new' water into the survey area.
- Effort has thus been expended on improved sensing of the ever-changing water column (depth, salinity, water velocity, surface barometric pressure).
- Clock drift in nodes has been improved over the years but in any case can be adequately determined.
- Battery life has improved from 28 days to around 110–180 days.
- Middle water-column node transfer was developed, and also high-spec node loaders, so that by 2017/18, a factor of 2 in survey efficiency had been gained.
- 80 nodes per day can be handled at a spacing of around 400 m.
- Emplaced optical fibre systems (DAS) are now in use for VSPs.
- Technology has been successfully developed to extend the low-frequency bandwidth.
- There is currently much research activity into AUV technology.
- Papers have appeared discussing 'swarms' of AUV/nodes.

- Continuous recording together with the known position of each and every seismic source operating, including 'seismic interference' might add additional useful information using simultaneous source technology.
- Automatic velocity model building might be possible with FWI.

References

Abma, R., Howe, D., Foster, M., et al., 2015. Independent simultaneous source acquisition and processing. *Geophysics,* **80**, WD37–WD44. DOI: 10.1190/geo2015-0078.1.

Beaudoin, G. and Michell, S., 2006. The Atlantis Project: OBS nodes: Defining the needs, selecting the technology and demonstrating the solution. Offshore Technology Paper 17977.

Beaudoin, G., Reasnor, M. D., Pfister, M. and Openshaw, G., 2010. First wide-azimuth time-lapse seismic acquisition using ocean bottom seismic nodes at Atlantis field-Gulf of Mexico. *Extended Abstracts, 72nd EAGE, Conference and Exhibition.*

Beaudoin, G. and Ross, A. A., 2007. Field design and operation of a novel deepwater, wide-azimuth node seismic survey. *The Leading Edge*, **26**(4), 494–503.

Boëlle, J.-L., Brechet, E., Ceragioli, E., et al., 2012. A large-scale validation of OBN technology for time-lapse studies through a pilot test, deep offshore Angola. *The Leading Edge*, **31**(4), 397–403.

Boëlle, J.-L., Ricarte, P. and Suiter, J., 2005. Sparse receiver and multi-azimuthal simulations from a high fold OBC campaign in the UK North Sea. In *SEG Abstracts*, 2005, 88.

Brenders, A., Dellinger, J., Chinaemerem, K., Qingsong, L. and Mitchell, S., 2018, The Wolfspar Field Trial: Results from a low-frequency seismic survey designed for FWI. *Expanded Abstracts*, *88th Annual International Meeting, SEG,* 1083–6.

Cole, R. A. and French, W. S. 1984. Three-dimensional marine seismic data acquisition using controlled streamer feathering. In *SEG Abstracts* 1984.

Dellinger, J., Ross, A., Meaux, D., et al., 2016. Wolfspar®, an FWI-friendly ultra-low-frequency marine seismic source. In *SEG International Exposition and 86th Annual Meeting*, 2016.

Dellinger, J., Brenders, A., Pool, R., et al., 2019, The Wolfspar® Field Trial: Testing a new paradigm for lowfrequency 3-D velocity surveys. In *81st EAGE Conference and Exhibition*, 3–6 June 2019, London.

Farhadiroushan, M., Parker, T., Shatalin, S., et al., 2019. Advanced geophysical measurement methods using engineered fiber optic acoustic sensor. In *81st EAGE Conference and Exhibition*. DOI: 10.3997/2214-4609.201901247.

Hampson, G., Stefani, J. and Herkenhoff, F., 2008. Acquisition using simultaneous sources. The Leading Edge, **27**(7), 918–23.

Hoffe, B. H., Cary, P. W. and Lines, L. R., 1999. A simple and robust method for combining dual-sensor OBC data. CREWES Research Report, Vol. 11.

Houbiers, M., Roste, T., Thompson, M., Szydlik, B., Traylen, T. and Hill, D., 2011.Marine Full-azimuth field trial at Heidrun revisited. In *SEG Annual Meeting*, San Antonio.

Houck, R. T. 2010. Reservoir. In *Methods and Applications in Reservoir Geophysics.* Tulsa, OK: Society of Exploration Geophysicists.

Ikelle, L. and Amundsen, L., 2005. *Introduction to Petroleum Seismology*. Tulsa, OK: Society of Exploration Geophysicists.

Johns, T. D., Vito, C., Clark, R. and Sarmiento, R., 2006. Multicomponent OBC (4C) prestack time imaging: Offshore Trinidad, Pamberi, LRL Block. In *SEG Annual Meeting*.

Johnston, D. H., 2010. *Methods and Applications in Reservoir Geophysics*. Tulsa, OK: Society of Exploration Geophysicists.

Keggin, J., Rietveld, W., Benson, M., et al., 2007. Multi-azimuth 3D provides robust improvements in Nile Delta seismic imaging. In *69th EAGE Conference and Exhibition*.

Kommedal, J. H., Fowler, S. and McGarrity, J., 2005. Improved P-wave imaging with 3D OBS data from the Clair field. *First Break*, **23**(12). DOI: 10.3997/1365-2397 .2005023.

MacLeod, M. K., Hanson, R. A. and Bell, C. R., 1999. The Alba Field ocean bottom cable seismic survey: Impact on development. *The Leading Edge*, 18(11), 1306–12.

Muyzert, E. 2018. Design, modelling and imaging of marine seismic swarm surveys. *Geophysical Prospecting*. DOI: 10.1111/1365-2478.12671. With further discussion by Gijs Vermeer in *Geophysical Prospecting*, **68**, 2020.

Naldrett, G., Soulas, S., van Gestel, J.-P. and Parker, T., 2020. First subsea DAS installation for deep water reservoir monitoring. In *First EAGE Workshop on Fibre Optic Sensing*, 9–11 March 2020, Amsterdam.

Parker, T., Shatalin, S. and Farhadiroushan, M., 2014, Distributed acoustic sensing: A new tool for seismic applications. *First Break* I32(2), February, 61–9. DOI: 10.3997/1365-2397.2013034.

Reasnor, M., Beaudoin, G., Pfister, M., et al., 2010. Atlantis time-lapse ocean bottom node survey: a project team's journey from acquisition through processing. In *SEG Abstracts* 2010.

Roende, H., Sheng, J., Liu, Z. and Bate, D., 2019. Considerations for a model building paradigm shift in the Gulf of Mexico. In *81st EAGE Conference and Exhibition*, June 2019, Vol. 2019, 1–5.

Sil, S., Srivastava, R. P. and Sen, M. K., 2009. Observation of azimuthal anisotropy on multicomponent Atlantis node seismic data. In *SEG International Exposition and Annual Meeting*, Houston.

Soubaras, R. and Dowle, R., 2010. Variable-depth streamer: A broadband marine solution. *First Break*, **28**(12), 89–96.

Stopin, P. J., Hatchell, P. J., Corcoran, C., Beal, E., Gutierrez, C. and Soto, G. 2011. First OBS to OBS time lapse results in the Mars Basin. In *SEG Annual Meeting*.

van Gestel, J.-P. and Anderson, G., 2017. Integration of time lapse seismic observations into the reservoir model: A case study on Atlantis. In *SEG International Exposition and 87th Annual Meeting*.

van Gestel, J.-P., Roberts, M., Davis, S. G. and Ariston P. O., 2013. Atlantis Ocean Bottom Nodes time-lapse observations. In *SEG Annual Meeting*, Houston. DOI: 10.1190/segam2013-1403.1.

Zhang, Q., Abma, R. and Ahmed, I., 2013. A marine node simultaneous source acquisition trial at Atlantis, Gulf of Mexico. In *SEG Annual Meeting, Houston*. DOI: 10.1190/segam2013-0699.1.

8

Microseismic Technology

PETER M. DUNCAN

8.1 Introduction

8.1.1 Overview

Microseismic technology as applied to the petroleum exploration and production industry finds its roots in classical seismology. In classical seismology the acoustic signals generated by earthquakes are captured and analysed to provide information on the location of the earthquake, the mechanics of the event and the nature of the earth along the path the signal traversed between the source location (hypocentre) and the seismic observation station. The motivations for that field of endeavour are public safety and a better understanding of the earth's structure and dynamics. While limited work has been done using earthquakes as seismic sources to create images suitable for hydrocarbon exploration, especially in areas where conventional seismic is prohibitively expensive, the majority of what is referred to as microseismic technology has made use of the microseismic events induced by production and development processes in the field, especially hydraulic fracturing or 'frac'ing', in order to monitor, understand the effects and control these processes to optimise hydrocarbon recovery from the reservoir. It is this frac monitoring application on which we will concentrate in this chapter.

8.1.2 History

Grechka and Heigl (2017) trace the industrial application of microseismic monitoring back as far as the early 1900s. Seismographs were deployed in mines to detect tremors that were often precursors to rock bursts and other safety hazards. The correlation of waste fluid injection at the Rocky Mountain Arsenal with an increase in local seismicity near Denver, Colorado in the mid-1960s (Evans, 1966) led to a series of injection experiments at Chevron's Rangeley, Colorado Field between 1969 and 1973. This established a relationship between injection pressures, fluid flow and induced seismicity (Raleigh et al., 1976). Grechka and Heigl go on to track the development of frac monitoring beginning with work done by the El Paso Natural Gas Company and the Sandia National Laboratory in the early 1970s through to the now famous Cotton Valley experiments of the late 1990s (e.g., Rutledge et al., 1998). It was in this latter work that much of the modern techniques for data capture and event location using downhole geophone arrays were established. The explosion of shale gas and shale oil exploration and development (the 'shale gale') beginning in the Barnett Formation of North Texas after 2003 led to an increase in frac'ing and the concurrent development of microseismic technology as it exists today.

8.2 Data Acquisition

8.2.1 Overview

Acquiring microseismic data is no different than the acquisition of conventional seismic data with the exception that it is a passive endeavour that requires no sound sources such as vibrators, airguns or dynamite. It does require the deployment of listening devices (geophones typically) and recording equipment of an appropriate bandwidth and sensitivity at positions suitable for the location and characterisation of the microseisms of interest. Current techniques can be grouped into three categories: downhole, surface and near surface or shallow buried arrays.

8.2.2 Downhole Arrays

Downhole monitoring is the legacy microseismic method. Currently it represents the largest portion of monitoring work done. The event magnitude (Mw) of the microseisms generated by oilfield processes such as frac'ing is typically in the range -3.0 to 0.0, that is two to five orders of magnitude less than the earthquake signals that seismologists usually study and well below what can be felt at the surface. A standalone three-component seismic observation station placed on the surface is unable to resolve microseismic events of this magnitude with a sufficient signal-to-noise ratio to locate the events with the needed precision and accuracy. An obvious solution is to place the geophone sensor closer to the event in a borehole drilled to the depth of the reservoir and near the treatment well. Ideally, one would like to have geophones at positions surrounding the events to facilitate event location with triangulation and trilateration. Generally speaking, it is rare that existing wells are available to provide this receiver geometry. Drilling new wells to accommodate microseismic monitoring is cost prohibitive. A half-step to making a single monitoring well more effective is to use an array of

Figure 8.1 A typical downhole array monitoring configuration consisting of a single vertical well offset from a pad of horizontal wells being stimulated. Event hypocentres are plotted as spheres along the lateral. The figure portrays the seismic signal travelling to the geophone array. The insert shows a hypothetical P (yellow) and S (blue) event seismogram as might be recorded.

multiple (10–100) three-component phones usually at a 25–50 m spacing along the wellbore (see Figures 8.1 and 8.2). These phones can be in the vertical or the horizontal portion of the well or both. Where possible, occupying multiple wells with arrays as in Figure 8.3 is to be preferred (Baig et al., 2010).

When suitable monitor wells are available the downhole monitoring methodology is quite easily executed. The advantages of the method are as follows:

- Small field effort. A crew of two can operate the wireline truck, deploy the array and record the data in a single monitor wellbore.
- Ease of permitting. Since the equipment is deployed in wells already likely owned by the operator there is no permitting required.
- Visible events. With the geophones located close to the events, a large number of the event signals are visible on the individual geophone field records lending confidence to the proceedings and a ready QC that the equipment is functioning.
- Real-time locations possible. With so few stations in the array it is possible to achieve reasonable hypocentre estimates for the larger events in real time as the data are collected.

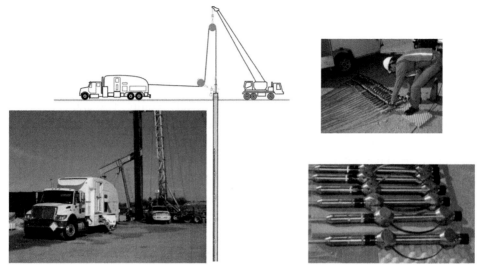

Figure 8.2 The various elements of a downhole monitoring field operation. The cartoon illustrates the wireline truck and crane configuration used to deploy the geophones downhole. The accompanying photographs show such a truck and crane as well as the geophones.

Figure 8.3 An illustration of a downhole monitoring operation with three vertical geophone arrays.

- Precise event locations. The high-frequency content of the signals recorded downhole enables precise event location, that is, the internal uncertainty is generally in the range of $\pm$ 10 m or less for the events within a few hundred metres of the monitor array.

The disadvantages of the technique include

- Monitor well(s) are required. The array must be placed at or near the reservoir depth in a well that is in suitable condition and appropriately positioned relative to the treatment well. It is cost prohibitive to drill new wells solely for the purpose of monitoring, so one must rely on there being an older well that can be made available to act as the monitor well. In all cases the monitor well(s) must be taken off production, production tubing removed, the well cleaned out and filled with fresh water and the casing/cement bonding must be in good condition.

- Location errors are large unless multiple wells are used. Baig et al. (2010) demonstrate with a theoretical dataset that one really requires two and preferably three monitor wells surrounding the stimulated events, as depicted in Figure 8.3 to achieve reasonable estimates of frac lengths and stimulated reservoir volume (SRV) (see Mayerhofer et al., 2010).
- Equipment is expensive and at risk. Downhole can be a harsh environment chemically, mechanically and from a temperature point of view. All of these factors present risks to the downhole equipment, especially when the kit is tractored into a lateral.
- Temperature limitations. In addition to the overheating risk, protecting the electronics in the downhole array may mean deploying the array shallower than would be ideal for the imaging geometry.
- Risk to the wellbore. Having the geophones get stuck in the well risks not only losing the geophones but also damaging the wellbore.
- Limited detection distance. The radius of investigation from a vertical array is typically 500–1,000 m (Maxwell et al., 2010). Beyond that distance the attenuation of the signal will reduce the useable bandwidth for arrival time picking (see Section 8.3) to the point where event location at the required precision is not possible. This detection distance is greater for larger magnitude events if the signal to background noise is larger.
- Limited imaging distance. In addition to seismic attenuation issues, the useful imaging distance from a downhole array is controlled by the aperture of the array. Eisner et al. (2009) show that the uncertainty related to the array geometry grows with distance from the array. Beyond a distance of about three to five times the array aperture, the improvement in precision as a result of having multiple geophones in the array tends to zero and the uncertainty in the estimated hypocentre becomes large.
- Multiple array deployments. Given the limited monitoring distances referenced earlier, monitoring a long lateral or a multiwell pad will require multiple repositioning of the monitoring array(s) in either the vertical or horizontal portion of the wells. This results in standby time for the frac crew, additional equipment risk, and the need to reorient the geophones in the array each time they are repositioned. A more efficient approach is to deploy multiple arrays as shown in Figure 8.3.
- Limits on moment tensor and mechanism estimation. The description of the focal mechanism of the event is critical to an analysis of the data that goes beyond

simple hypocentre estimation. Vavryčuk (2007) has shown how having a single vertical array in a monitor well makes it impossible to completely determine the moment tensor of the event.

8.2.3 Surface Arrays

As shale development began to expand after 2003, field development moved into new areas where monitor wells were not available and into areas such as the Haynesville of southwestern Louisiana where the temperatures at reservoir depth were too high for existing downhole tool technology (Duncan et al., 2013). The monitoring solution adopted was to deploy relatively large areal arrays of conventional 1-C (single component) geophones on the surface. Once deployed the array remained in place for the duration of the treatment. After experimentation with rectangular arrays, an array of the form depicted in Figure 8.4 became the most common surface configuration.

This circular array typically consists of 10–20 arms, each with a length approximately equal to the depth to the treatment zone. Each arm has receiver stations equally spaced along its length at an interval of 12.5–50 m depending on the depth to target. This results in the total number of stations in such an array to be 1,000–4,000. Each station will have 6–12 geophones within its group. These geophones are planted regularly along the arms over the interval between stations. A circular array is appropriate for monitoring a vertical treatment well and will be positioned symmetrically about the wellhead. For the case of a horizontal well, the arms along the direction of the lateral will be extended to preserve the horizontal

Figure 8.4 Illustration of a surface monitoring array. The 11 blue lines represent the strings of geophones laid out radially from the wellhead. Each line or arm would typically have 100–200 geophone groups consisting of 6–12 vertical one-component geophones connected in series.

Figure 8.5 Map of a real surface monitoring array superposed upon the local topographic sheet. The locations of the five horizontal wells being monitored are shown as red lines. The black lines are the arms of the array. Note how some lines are bent and displaced owing to permitting issues or other obstructions. Also note how the arm length is increased for the arms above the laterals. For reference, the laterals in these wells are about 1,500 m (5,000 ft) long.

offset between the receivers and the events (see Figure 8.5). The configuration can be expanded easily to cover laterals of several thousand meters length as well as multiwell pads.

Some of the advantages of the surface array methodology over downhole techniques are as follows:

- No monitor well(s). The result is a reduction in cost and risk.
- Expandable. By lengthening the arms, the setup can accommodate long laterals and multiwell pads without movement of the array once deployed.
- Location uncertainty is constant over the image volume. Since the array is essentially a constant distance from the events, the uncertainty as a result of array geometry is effectively constant.
- Excellent focal mechanism detection. Since the array captures a large fraction of the focal sphere, it is possible to make a good estimate of the moment tensor even if only P-waves are recorded (Eaton and Forouhideh, 2011).
- Limited intrusion on operations. With a surface array monitoring, no wells need be taken off production nor prepared for use as a monitor well. The monitoring operations are off the treatment pad, so they do not interfere with the treatment process.
- Field equipment is not at great risk. The surface methodology uses conventional seismic geophones and recorders that are deployed in areas where they are not likely to suffer significant damage or outright loss. Since the same equipment is

used in many seismic applications it is relatively inexpensive compared to downhole geophones.

Some disadvantages of the surface array technique are

- Large field effort. To deploy several thousand geophones over an area measuring tens of square kilometres will take a crew of between 15 and 30 people several days depending upon the terrain, vegetation and the degree to which the area has been developed.
- Permitting required. The areal extent of the array makes it likely that several different surface landowners will be traversed by the array. Permission must be obtained from each and often a payment to obtain access is required.
- Signals are not visible in raw field data. Since the geophones are now several thousand meters above the target reservoir and planted on what is likely a noisy surface (especially if there are production operations in the vicinity), the event signals are mostly hidden in the noise in the raw field data. It requires the power of the stack of a thousand channels or more to bring these signals out of the noise. This can be disconcerting in the field as one is unsure if anything useful is being recorded.
- Real-time analysis is more costly. Owing to the size of the data set being collected, event location and higher order imaging in real time requires expensive computing equipment to be deployed in the field given the communication capacities available at the time of this writing.
- Higher magnitude detection threshold. Typically, a surface array will have a magnitude of completeness about half a magnitude unit higher than will a downhole array owing to the natural attenuation of the signal along its travel path and the higher ambient noise level at the surface.

8.2.4 Buried Arrays

The large geophone count (high fold recording) used in surface arrays is driven by the need to overcome the high levels of ambient noise through the stacking process. Much of this noise disappears at relatively shallow depth beneath the earth's surface. In the buried array method of monitoring (see Figure 8.6a) the geophones are placed in shallow boreholes beneath the surface. The phones are typically cemented in and become a permanent, life-of-field monitoring facility. Since the ambient noise level is reduced, a lower fold of observation is required to achieve the same signal-to-noise ratio in the observations. The rate that the noise falls off with depth tends to decrease after the first few 10's of meters as shown in Figure 8.7 (Duncan et al., 2013). The cost of drilling each station's borehole obviously increases with depth. There is an economically driven trade-off between

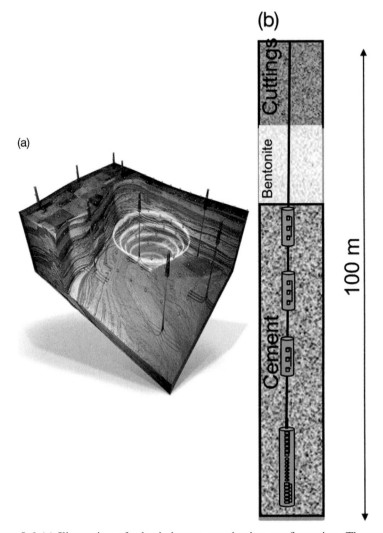

Figure 8.6 (a) Illustration of a buried array monitoring configuration. The orange cylinders in the block diagram represent a hypothetical array layout. In reality, the stations are not likely to be quite so regularly distributed owing to permitting and access concerns. (b) Illustration of the configuration of each station. In this case a three-component geophone consisting of six elements in each component direction is planned for the deepest geophone location. Above that are placed three vertical single component phones consisting of three elements each.

having more stations at shallower depth or fewer stations placed at greater depth. In practice these arrays typically have geophone depths in the 10–100 m range with two to four stations per square kilometre. At the deeper end of this range, it is often useful to put more than one level of geophones in the borehole at

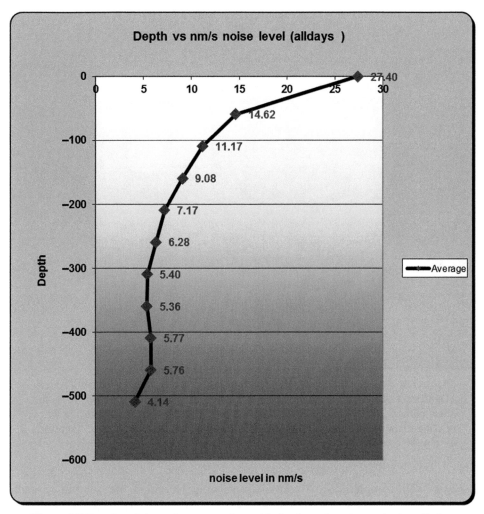

Figure 8.7 Results of the noise attenuation with depth study prior to the installation of the Encana Blacksmith Buried Array in southern Louisiana. The rms noise measure on the horizontal axis is computed in 60-second windows, averaged over 3 days of data recording. The depth of the observation geophones below surface in feet is on the vertical axis. The noise measures are annotated on the curve. Based on these results the geophones in the Blacksmith array were placed at a 250-ft depth (Duncan et al., 2013).

approximately 15 m intervals as depicted in Figure 8.6b. Geophones may be single vertical component or 3-C (three-component).

Once the geophone stations have been installed there are two modes of operation available: discontinuous and continuous. In the discontinuous mode the seismic recorders are deployed temporarily on an as needed basis to monitor

Figure 8.8 This picture shows a single station of the Encana Blacksmith array (Duncan et al., 2013). The array was used in a discontinuous fashion. Note the three autonomous, single seismic channel recoding nodes connected to the wires leading to the buried 3-C geophone. An automotive battery is supplying the power to run the recorders.

specific well completions. Autonomous seismic recorders are set out at each station for the monitoring period as pictured in Figure 8.8. The data are collected manually at regular intervals of a few days or perhaps weeks. For continuous monitoring, a more permanent installation of the seismic recorders is made (see Figure 8.9). Data are collected by an appropriate telemetry system such as Wi-Fi or a cell phone network. This automated data collection enables real-time data analysis to be performed.

Buried arrays are designed to give similar event imaging results to surface arrays and have all of the same advantages compared to downhole methods as listed earlier for surface arrays. In addition, the buried array approach has some advantages over surface arrays as follows:

- Reduced field effort. Once installed the buried array needs only a small crew to function. In the discontinuous mode a crew of two to six is needed depending

Figure 8.9 A single buried array station from an array in south Texas configured for continuous operation and real-time analysis. The seismic recorder and communications kit are in the locked box on the mast. Note the solar panel and small wind turbine supplying power to keep an automotive battery in the box charged. The antennae on the top of the mast allow for system control and transmission of the data to a central location.

upon how often the data are collected. In the continuous mode of operation, the array can be essentially un-manned except for occasional maintenance.

- Reduced permitting costs. Since the stations are far apart, not connected by any wires, and can be placed in an almost random fashion, it is usually easy to find places to install the station at a reasonable permitting cost.
- Life of field asset. The geophones in the ground are in a very benign environment and, if left undisturbed, will remain operational for many years. The array can be used again and again over the life of the field. It can also be used for acoustic reservoir monitoring and as a permanent receiver for 4D active seismic imaging.
- Lower unit monitoring costs. Owing to the reduced crew size and reduced number of receiver stations, the cost of buried array monitoring is typically less than for a surface array if the cost of drilling and installing the array is amortised over a sufficient number of wells. The unit cost continues to decrease the more the array is utilised.
- Large coverage area. The array can be expanded easily to cover an entire field and, in fact, the costs can be shared by adjacent operators using different parts of the same array.

- Lower detection threshold. Owing to the lower noise environment, a buried array will typically push the detection threshold of events lower than a surface array in the same location.

There are of course some disadvantages to buried arrays as well, such as

- Drilling. This is required as part of the installation process.
- Costs are front loaded. The biggest single cost is for the drilling and initial installation of the buried array and attendant equipment. After installation the incremental cost for monitoring additional wells is small.
- Permitting. These costs are ongoing (usually annual) if the array is left in place.
- Removal/remediation is a requirement. When the array is no longer to be utilised, it is usually necessary to remove the surface equipment, cut the wires to the buried geophones and plug the boreholes to near the surface.
- Real-time monitoring is more difficult. A wireless telemetry system is required to gather the data and transmit it to a processing facility. This means an additional expenditure.
- Receiver notching and surface ghosting issues. Buried phones may be subject to spectral notching resulting from the surface reflection. Surface arrays also experience the signal enhancement of the free surface ghost (Cieslewicz and Lawton, 1996) which is lost when the phones are placed at depth.

8.2.5 *Hybrid Arrays and Distributed Acoustic Sensing*

The three array styles discussed earlier represent end members of the acquisition approaches available. Downhole and surface arrays are sometimes deployed simultaneously to test one against the other or to take advantage of the strengths of each. A small array of 10–20 stations deployed on the surface over a completion being monitored with a downhole array can deliver information on event focal mechanisms that would otherwise be lost. Similarly, a single group of geophones planted at the surface of each buried array station has been found to increase the imaging capability of the array.

Distributed acoustic sensing (DAS) using fibre-optic cable is a relatively new development that is being considered as a replacement for downhole geophones. DAS, in temporary installations, sometimes called intervention fibre placement, has the advantage of being relatively inexpensive and less risky than placing geophones into the horizontal well with a tractoring device. The large aperture of observation is a great improvement over a string of geophones. However, the 1-C nature of DAS sensing is an issue for microseismic imaging applications that is most easily overcome by once again having multiple monitoring wells with a fibre in each (Chambers, 2019).

8.3 Event Hypocentre Estimation

8.3.1 Basics

In the previous section we saw that there are two prevalent methods of acquiring microseismic data: downhole and surface. What drives this dichotomy is that there are two prevalent approaches to locating microseismic events: single station arrival time picking and full-waveform migration sometimes called beam steering. The former requires a broadband signal at the receiver such that the arrival time of the P and S phases of the signal may be identified or picked with an accuracy on the order of 1 ms. The difference in the arrival times gives an estimate of the distance of the hypocentre from the receiver under the assumption of a reasonable velocity model. If a 3-C geophone is used, then the orientation of the arriving wavefront can be measured giving an estimate of the bearing of the hypocentre from the receiver (Lay and Wallace, 1995). It is the signal bandwidth necessary to implement this location methodology that drives the deployment of the geophones downhole close to the events so that high-quality observations of both the P and S phase arrival times are acquired.

The migration approach uses the full recorded waveform, rather than just the picked arrival times. The location in the subsurface that results in the best statistical alignment of the waveforms recorded at each station after accounting for the time delay and filtering that the signals experience along their individual travel paths is taken as the event location estimate (Duncan and Eisner, 2010). Note that this migration methodology also requires a reasonably accurate velocity model and the ability to forward model the wave propagation.

In fact, either location approach could be used with either acquisition approach. Practically speaking, the downhole method generally has too few stations over too small an aperture to give low uncertainty results using a full-waveform migration approach. This is precisely why the arrival times are picked and the waveforms are then discarded. It is perhaps insightful to recognise that, after picking, the process of location reduces to a migration of the noise free arrival time picks. On the other hand, geophones on or near the surface generally have too narrow a useable bandwidth to allow for picking of arrival times with sufficient accuracy. As well, near surface affects make observation of the shear wave component of the signal very problematic so arrival time differences between the phases are not available.

The uncertainty in the hypocentre estimates in either of the aforementioned methodologies will depend upon the geometry of the acquisition array relative to the event location (Eisner et al., 2009) and the useful bandwidth of the data. The bigger the apparent aperture the less uncertainty in the hypocentre estimate. The higher the signal-to-noise ratio of the recorded traces the smaller the uncertainty in the hypocentre estimate. The accuracy of the hypocentre estimates, as opposed to uncertainty, will depend on the accuracy to which the acquisition geometry is known and the accuracy of the velocity model. If the monitor well or receiver stations are misplaced in the geometry setup, the location estimates will be wrong (Bulant et al., 2007). If the velocity model is incorrect then the travel time model will be systemically wrong, and all events will be mis-located. Either methodology benefits from having controlled events such as string shots or perforation shots set off in the treatment well at known locations and times to calibrate the velocity model. Taking into account anisotropy is important in modelling shale velocities (Gretcka and Heigl, 2017)

8.3.2 Pros and Cons of Arrival Time Picking

The positive attributes of the arrival time picking methodology include

- Small computational effort. Locations of visible (pickable) events can be estimated with as little as a single 3-C station. Processing effort is small and real-time location estimates are very feasible at least for the larger event magnitudes.
- Hands on data interpretation. An experienced picker can 'see' through the noise, often producing passable results even for low signal-to-noise records.
- Focal mechanism independent. The location estimate is independent of knowing the focal mechanism.

The drawbacks to the arrival time picking methodology are

- Picking arrival times is not easily automated. Auto pickers do exist but most seismologists still rely on hand picking to arrive at final location estimates.
- Labour intensive. Manual picking of thousands of events is a rather large task.
- Events must be 'visible' on individual traces to be picked. This results in limited detection distance.
- Subjectivity. Each picker tends to interpret arrival times in their own way. If a single project has multiple pickers, then there can be differences between the subset of events each picks.
- Real-time failure. Since real-time results are usually machine picked and post processing results are hand-picked, there is often a large discrepancy between the early and final results leading to distrust in the data in general and an unwillingness to use the data as a guide for adjusting treatment parameters on the fly.
- Velocity model sensitivity. The method is very sensitive to velocity model errors, to the velocity structure and to the treatment of anisotropy.

8.3.3 *Pros and Cons of Full-Waveform Migration*

The advantages of the full-waveform migration methodology are

- Easily automated. There are many well understood techniques for migrating multitrace seismic data.
- Picker independent. As no human picking is done, the problem of subjectivity is removed.
- Low signal magnitude detection threshold. The power of the stack is at work here.
- Scalable. Since the computer is doing the work, it is possible to have large areal surface arrays and large subsurface coverage. This is useful for long lateral, multiwell, multipad monitoring at reasonable cost.
- Less sensitivity to velocity model error. The long travel paths typical of data processed in this method tend to average out velocity model errors.

Some of the disadvantages of the beam steering methodology are

- Large fold and aperture are required. Many stations and hence potentially expensive field effort become necessary.
- Large processing effort. We are basically depth migrating a long time, multi-channel data volume. Computer utilisation is high.
- Focal mechanism is required. A phase correction for an assumed focal mechanism must be applied to the data prior to the beam steering process or an incorrect location estimate will be made. The need to determine focal mechanism could be seen as an opportunity rather than a disadvantage.

8.3.4 *Costs*

It is understandably difficult to be very definitive about monitoring costs when the number of variables to be considered is so very great. A few general comments are possible. Acquisition of the data is usually about 75% of the cost of a monitoring project. Acquisition of downhole or surface array data is driven by the number of crew days the crew is deployed. At the time of this writing, a single wireline crew will be charged at around $10,000/day. Tractoring devices to drag the wireline along a lateral can be as much as $50,000/day. An average surface array crew will be charged at $25,000/day. Mobilisation, demobilisation and standby days for weather or other causes will be often at a reduced rate. At these rates monitoring a vertical well with a single downhole array will be on the order of $50,000–$100,000 exclusive of data analysis. Please note, this cost does not take into account the cost of sourcing and preparing a suitable monitor well. The same well monitored with a large surface array might be double that cost for the crew but would not incur any well preparation costs. On the other hand, a four-well pad of horizontal wells would likely cost on the order of $500,000 to monitor with a surface array and be at least that much if not more for a downhole crew, especially if multiple monitor wells and/or tractoring are involved.

At the time of this writing an installed buried array station, when a sufficient number of stations are ordered, costs about $8,000–$12,000 depending on the depth of installation. Seismic recording and telemetry equipment to enable continuous real-time monitoring will cost an additional $7,000–$10,000/station. An average minimum array will require 100 stations. The small crew to acquire data discontinuously for a few wells will cost in the range of $4,000–$8,000/ crew day. In a permanent system with real-time monitoring capabilities the operating expense is one-third to one-half that amount.

Another useful way to look at expenses is on a cost per stage basis. Assuming a reasonable minimum size job over which one can amortise mobilisation and setup costs, a typical downhole monitoring project using a single monitor well will cost in the range of $10,000–$20,000 per stage through event location analysis. A representative surface array project will cost in the range $8,000–$15,000 per stage through hypocentre estimation. A buried array project can cost as low as $5,000/stage if enough wells are monitored over the life of the array. The higher order analysis that is described in the next sections will typically increase the costs by 10%–25% depending on how far the analysis is taken.

8.4 Beyond the Dots-in-the-Box

8.4.1 Overview

Images of the hypocentre distribution around the treated wells such as is shown in Figure 8.10 have been the primary tool for presenting and interpreting microseismic results with regard to treatment effectiveness. The distribution of events is used to provide answers to such questions as

- What is the local direction of maximum horizontal stress as evidenced by the average azimuth of the hypocentre trend?
- What frac length was achieved?
- What frac height was achieved?
- What stimulated reservoir volume was achieved?
- Did the treatment stay withing the target reservoir zone?
- Did the treatment encroach on adjacent wells?
- What should the well spacing be?
- Did the treatment from adjacent stages overlap?
- What should the stage spacing be?

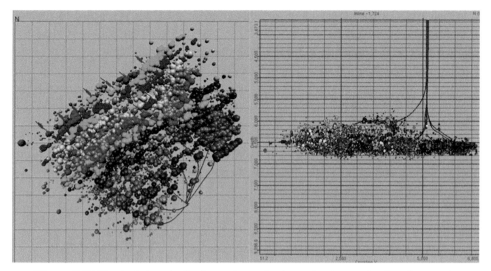

Figure 8.10 Typical dots-in-the-box view of microseismic monitoring data; map view on the left and section view on the right. Treated wells are represented by blue lines. Hypocentres are represented as spheres whose size is proportional to magnitude. Events associated with a given frac stage have a common colour on each well. The grid cells on the map view are 500 × 500 feet (152 × 152 m).

- Were there changes in reservoir response to treatment along the well?
- If the treatment parameters were varied, how did the microseismic data change in response to those differences?
- If diverters were deployed, were they effective?
- Did the frac excite movement on regional faults?
- Was there any evidence of casing failure, bad cementing of the well, plug failure or sleeves not performing as planned?

Another useful display of these hypocentre locations is in animated form. Movies depicting the growth of the microseismic volume as the treatment progresses allow for interesting insights into how the reservoir responded to the treatment. Usually these movies will display the pumping program (instantaneous pressure, slurry rate and proppant concentration) concurrently so that the appearance of events may be visually correlated with the progress of the treatment.

While the information provided by the dots-in-the-box is useful, it is clear that there are other important questions that need to be answered such as

- How is the proppant distributed within the stimulated reservoir volume?
- Can the microseismic data give insight on production rate and the expected ultimate recovery (EUR)?

- Which parts of the reservoir or which treatment program gives rise to greater hydrocarbon recovery?
- What will be the drainage volume of the treated wells over time?
- What is the optimum development scenario?
- What is the stress distribution near the wellbore as it relates to treatment, production and wellbore stability?

It turns out that microseismic monitoring can contribute to answering these questions as well. The key is to go beyond simply locating the events in time and space and to estimate the nature of the failure that created the observed signals, that is, the focal mechanisms of the events.

8.4.2 Radiation Pattern, Focal Mechanism and Fracture Modelling

Recognising that the observed microseismic signals are the result of failure and movement of rocks along faults and fractures in the subsurface opens the opportunity for a higher order analysis and modelling of the SRV. The nature of the rock movement (the focal mechanism) gives rise to a certain radiation pattern of the seismic signal. If one can sample the radiation pattern sufficiently it is possible to estimate the focal mechanism. Knowing the focal mechanism, we can create a model of the fracture and the movement that occurred on that fracture. This inversion to a fracture model is non-unique but with certain reasonable assumptions a useful modelling of the ensemble of the fractures that constitute the SRV is possible.

To invert the seismic data for the focal mechanism we need a model that predicts the seismic wavefield for a given failure event. One of the earliest models is due to Brune (1970) He calculates the displacement spectrum of the seismic signal for a modelled event assuming an instantaneous slip of a circular fracture of a given radius. Figure 8.11 is an example of such a spectrum and a Brune model fit to the observations (Baig and Urbancic, 2010). The low frequency plateau or spectral level can be used to estimate the moment magnitude (M_0) of the event. The corner frequency (f_c) predicts the radius of the area of instantaneous slip that has occurred (Maxwell, 2014). The value of M_0 can be related to the area of a planar failure element through the equation

$$M_0 = \mu S d \tag{8.1}$$

where μ = rock rigidity, S = fracture area and d = displacement. The product Sd is also known as the seismic potency. Following Williams-Stroud (2008) and Williams-Stroud and Eisner (2009) we can use Eq. (8.1) to estimate the area of planar elements whose failure resulted in the recorded seismic signals.

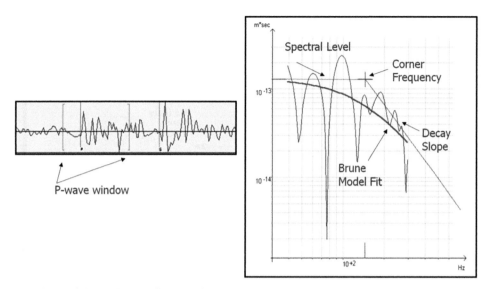

Figure 8.11 A P waveform and its displacement spectrum based on the shown P-wave window. The magnitude is calculated assuming the Brune model to determine the spectral level and corner frequency (Baig and Urbancic, 2010).

A complete description of the focal mechanism also requires knowing the strike and dip of the planar element as well as the rake or direction within the plane of the fracture of the ground motion. In Figure 8.12 the 'beach ball' representations of the focal mechanisms for certain common fault types are shown. Fortunately, the slip of the fracture results in a radiation pattern for the wavefront initiated by the event. Figure 8.13a illustrates the P-wave radiation pattern for a strike-slip fault. If the microseismic data collection is sufficient to characterise the observed radiation pattern, then one can use these data to estimate the remaining attributes of the focal mechanism. Figure 8.13b demonstrates how the radiation pattern shown in Figure 8.13a is sampled by a large surface array allowing for a characterisation of the focal mechanism.

Quite often a simple cataloguing and display of the estimated focal mechanisms offers important insights into the nature of the reservoir stimulation. Consider the microseismic results of Figure 8.14. The familiar dots-in-the-box display represents the hypocentre locations of the detected microseismic events captured during the treatment of four wells. The wells are drilled into the same reservoir, two from north-west to south-east, and two from south-east to north-west. The wells were completed at the same time. It is obvious that something different is happening mid-well where the treatment is seen to take off to the north-east. The cause for this is a matter for speculation.

Now consider the same dataset, but with the focal mechanism style of each event depicted through colour (see Figure 8.15). Red events are dominantly

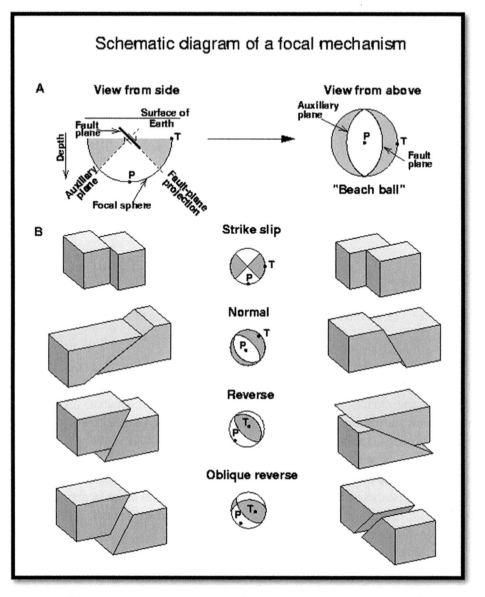

Figure 8.12 Focal mechanism representations (lower hemisphere beach balls) for common fault types (USGS, 1996 public domain). The fault plane occurs along the boundary of the compressive (P) and tensile (T) stress quadrants. The seismic radiation pattern alone cannot distinguish between movement along the fault plane or the auxiliary plane. Part A details the elements of the beach ball representation in section and map view. Part B illustrates the beach ball representation for various common fault styles displayed as block diagrams. The alternative and indistinguishable failure planes are shown.

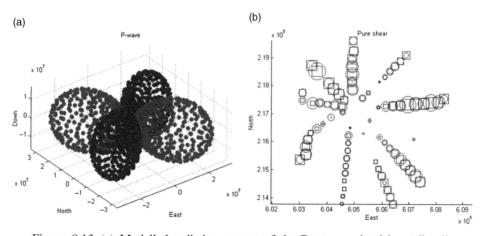

Figure 8.13 (a) Modelled radiation pattern of the P-wave emitted by strike-slip motion along a vertical fracture with either a north or an east strike and a horizontal rake. Blue represents the tensile stress quadrant. Red represents the compressive stress quadrant. (b) The seismic signal recorded by a star shaped microseismic surface array for a strike-slip event such as that in part (a). Circles represent the amplitude of the signal recorded. Squares represent the amplitude of the model found to fit the data. Blue signals are positive. Red signals are negative. (Courtesy of Leo Eisner)

dip-slip with a nearly vertical rake. Blue events are dominantly strike-slip with a nearly horizontal rake. Note that the blue events tend to be larger in magnitude (see Wessels et al., 2011). It is the author's experience that the reactivation of pre-existing regional faults during a stimulation usually results in larger strike-slip fault events that can occur at significant offsets from the treated well. It is also our experience that it is the reactivation of such pre-existing faults in a strike-slip mode that gives rise to wellbore damage during stimulation.

A more general approach to modelling the source mechanism of a point source acting across a planar surface is as nine force couples acting on that point (Aki and Richards, 2002). These are represented by the elements of the moment tensor as given in Eq. (8.2). Symmetry dictates that only six of these are independent.

$$
\mathbf{M} = M_0 \begin{bmatrix} M_{11} & M_{12} & M_{13} \\ M_{21} & M_{22} & M_{23} \\ M_{31} & M_{32} & M_{33} \end{bmatrix} \tag{8.2}
$$

The elements of the moment tensor may be estimated from a sufficient areal sampling of the amplitude and phase of the event displacement field. The process of such a moment tensor inversion (MTI) is well described in Grechka and Heigl (2017). Successful MTI allows for a complete description of the focal mechanism including the magnitude, strike, dip and rake. However, as Grechka and Heigl point out, this process can be unstable in the presence of seismic noise in the data and requires a reasonably accurate knowledge of the velocity field to be valid.

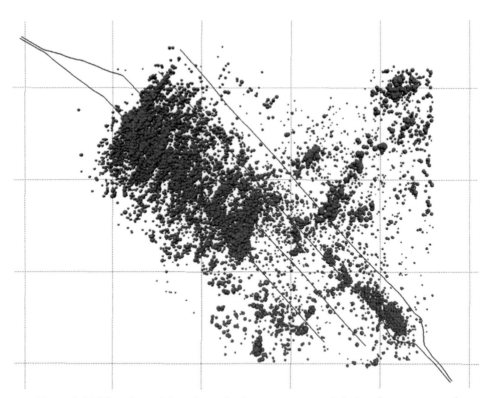

Figure 8.14 Map view of the microseismic events captured during the treatment of four wells (blue lines) drilled into the same reservoir. Circle size is proportional to event magnitude. Note the excursion to the north-east of the microseismic event pattern in the mid-section of the wells. The grid cells in the map are 1,000 × 1,000 feet (305 × 305 m).

With an estimate of the area of the fracture plane as well as its strike and dip, it is straightforward to now replace all the hypocentres in the frac model with planar elements, whose sizes relate to the moment magnitude of the events and whose attitudes are determined from the estimate of strike and dip (Williams and Eisner, 2009; Eisner et al., 2010; Williams-Stroud et al., 2010). As an example of this process consider the microseismic hypocentre plot presented in Figure 8.16. The microseismic monitoring data were acquired with a surface array consisting of 2,345 stations spread over 18 arms as depicted in Figure 8.17. The asymmetric layout was the result of a river channel on the north-east that presented a physical barrier to layout. Receiver station spacing along the arms was 25 m. Each station had a 25 m long, 12 geophone linear array evenly spaced along the arm.

Wells C and D are drilled into the Vaca Muerta Formation in the Neuquén Basin, Argentina (Crovetto et al., 2020). The monitored wells had an average depth over the treatment interval of 2,669 m subsea. The wells were zipper frac'd meaning that treatment alternated between wells with each stage in well

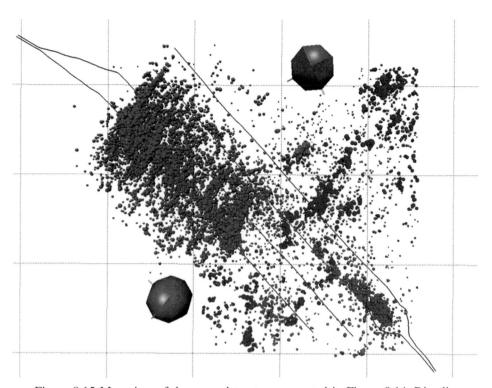

Figure 8.15 Map view of the same dataset as presented in Figure 8.14. Dip-slip events are coloured red. Strike-slip events are coloured blue. The strike-slip events are interpreted to represent the activation of a pre-existing regional fault set with a north-east strike.

D followed by a treatment of the corresponding stage in Well C. Well A, 750 m to the west, had been completed several months earlier.

The areal extent of the surface array allowed for an estimate to be made of the focal mechanism of each event using the methodology described above. These focal mechanisms were predominantly dip-slip, striking in a north-east to east–north-east direction (see Figure 8.18). A few dip-slip events with a strike of north–north-west were also observed as were some oblique- and strike-slip events generally striking east-north-east.

The observed magnitude, azimuth and dip of the frac events were used to replace each event with a planar representation of the fracture centred on the hypocentre. Taken together, these planar elements form a discrete fracture network (DFN) that describes the total SRV in a geologically insightful manner. The DFN for this project is pictured in Figure 8.19.

In addition to illustrating the derivation of a DFN description of the treatment, these data also illustrate an interesting frac-driven interaction (FDI) between the

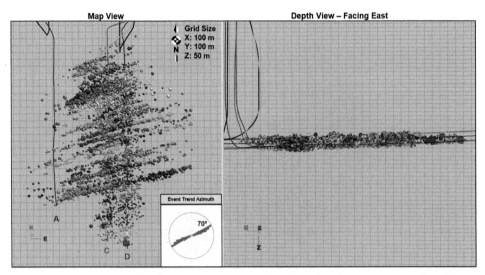

Figure 8.16 Microseismic event hypocentre locations for the treatments of wells C and D. Events are colored by stage and sized by magnitude. The rose diagram insert shows the average azimuthal trend of the events.

Figure 8.17 Layout of surface microseismic surface array. The red line represents well C. The purple line to the east represents well D.

treated wells and well A. Well A had been on production for approximately 7 months prior to the completion of the wells C and D and was on production when the treatment began. When the completion of wells C and D advanced to the point that the treating stage was on strike with the dominant fracture direction

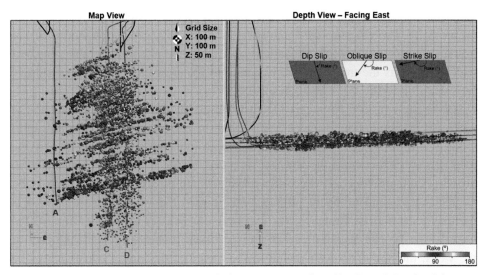

Figure 8.18 Event hypocentres of the treatment of wells C and D sized by magnitude and coloured by rake.

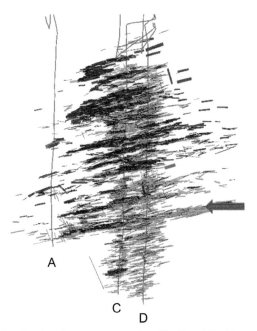

Figure 8.19 Total DFN for the treatment of wells C and D. Note the effect on the SRV of well A which was producing at the time of treatment. The red arrow indicates the timing of the shutting-in of well A during the treatment.

extended to well A, the nature of the fractures changed dramatically as seen in Figure 8.19. The fracture half lengths became much larger and extended to the primary well (well A) while remaining reasonably symmetric about the treatment wells. It appears the treatment had broken into fractures that had been previously activated during the treatment of well A.

Interestingly, at the point this FDI was observed, the pressure in the primary well, still on production, dropped. This is not the expected reaction as one would predict the inflow of frac fluid to increase the pressure in well A. One interpretation is that when the new frac's first connected with the drainage network of well A, the primary well fracture system was temporarily clogged, perhaps by having the original proppant pushed back toward the well. Thus, well A was partially cut off from the reservoir. The well was shut in for the duration of the treatment at approximately that time. As treatment proceeded toward the heel of the wells, the new fractures became less symmetric, showing a bias towards the primary well to the west suggesting communication with well A. When well A was brought back onto production following the treatment, there was a large jump in water production for a short time and a small amount of sand was recovered probably confirming some fluid communication between the wells.

The frac'ing of the Vaca Muerta shale oil play in Argentina's Neuquén Basin is further discussed in the case study presented in Section 8.6.1.

8.4.3 Modelling Production

At this point we have produced a more 'geological looking' model of the fractures created by the stimulation, but it is questionable what additional value this new model provides. Real value can be derived by using this model as input to a reservoir simulation which can then predict the production from the well, compare the production from wells treated with different procedures and perhaps, more importantly, predict the volume of the reservoir around the well that will be drained by the well over time. This last parameter is a primary driver in well spacing decisions.

An early example of this workflow was published by Kashikar et al. (2015). Their workflow progresses as follows:

- Following the method of Shapiro et al. (2002), microseismic events that occur at distances too far from the treatment point and that are too soon in the treatment time reasonably to be caused by fluid diffusion into the rock are designated as 'dry' events and removed from the model. It is recognised that there may be dry events inside what Shapiro terms the fracture front, but there is no recourse but to leave these in the model.

- A discrete fracture network (DFN) model of the frac is constructed in the manner described by Williams-Stroud (2008). The volume of rock containing the DFN is termed the stimulated reservoir volume (SRV).
- Following upon the method described by Neuhaus et al. (2014), that portion of the SRV that is likely propped during the treatment is estimated using a material balance approach. In this approach, the actual volume of proppant pumped in each stage of the frac is algorithmically placed into the DFN volume, using an assumed packing factor, from the treatment point outward until the proppant volume is exhausted. The portion of the SRV that is 'propped' is termed the productive SRV (P-SRV).
- Following the method of Oda (1985), the DFN is used to calculate a permeability enhancement value in each element of a finite element volume constructed to model the SRV. Other properties of the reservoir required to execute a reservoir production simulation for the treated well(s) are obtained from well logs, public databases or the operator's experience in the play.
- Reservoir simulation of the early production (typically the first 90 days) is used to calibrate the reservoir model (Khodabakshnejad et al., 2019). After this, longer time periods of production are simulated to estimate hydrocarbon and water production volumes as well as drainage volumes over time. The model can be used to predict the effect on production from individual wells for different well spacings to inform the well spacing decision.

8.4.4 What Is 'Optimum'?

The monitoring and modelling work described so far are designed to help the operator make decisions on treatment and well spacing parameters that optimise the development of the resource. It is important to be clear on the definition of 'optimum'. To the completion engineer the word 'optimum' might mean that he 'puts away' the prescribed treatment on time and under budget. To the microseismic monitoring geophysicist the 'optimum' well might mean the one that has the most events or perhaps has the most events constrained to the reservoir interval. To the production engineer the 'optimum' well might be the one with the highest initial production (IP) rate or the largest EUR.

In the end, none of these are particularly good definitions of 'optimum' because we should really consider the return on investment (ROI), not just the total revenue received or the total cost incurred. If one puts away the treatment but sands out the well resulting in little or no production, that is failure. If one increases the EUR for a given well but at a cost greater than the value of the incremental hydrocarbon production, that is failure. It really is necessary to consider full-cycle economics to judge what is optimal (Cao et al., 2017; Xiong et al., 2019).

(a)

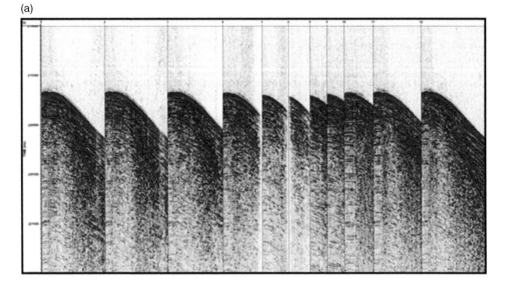

Figure 8.20 Illustration of workflow taking hypocentre distribution into a DFN model of the fractures created by the treatment. (a) Raw microseismic data. Each segment represents the offset ordered time gather of an event arrival along an arm of the geophone array similar to that shown in Figure 8.5. (b) The hypocentre distribution along the treated wellbore. Events are coloured by individual treatment stage. (c) The DFN derived from the hypocentre distribution following the method described here. (Data courtesy of Element Petroleum)

Consider the following example to illustrate this point. An operator in the Permian Basin of Texas completed a well in the well-known San Andres formation (Carlson et al., 1965) in 2018. The well stimulation was monitored with a surface microseismic array as described earlier. Figure 8.20 illustrates the progression of the analysis of these data from hypocentres through DFN. Figure 8.21 then compares the raw SRV derived from this DFN with the estimated propped portion of this same SRV, the P-SRV. Using these models, a series of reservoir production simulations was run for different development scenarios involving different numbers of wells within the same lease unit. The models were created by replicating the original SRV at different inter-well spacings to achieve the desired number of wells. The commonsense prediction is that the produced hydrocarbons will increase with the number of wells, but at some point, the wells will begin to interfere with each other, and the incremental production will not increase linearly with the number of wells.

Figure 8.22 shows the production estimate for a four-well and a six-well development scenario. As expected, the production increases with the number of

(b) (c)

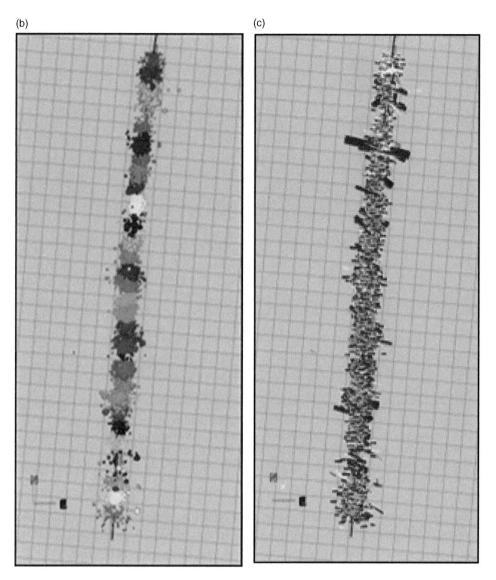

Figure 8.20 (*cont.*)

wells, in this case the 50% increase in the well count resulted in an almost 50% cumulative increase in hydrocarbon production. However, this increase in production resulted in extra drilling, completion and operating costs which, when properly accounted for, predict that the six-well scenario would result in a 28% improvement in net present value (NPV) for the full-cycle project. Much as reported by Cao et al. (2017), when models were run with a denser well spacing the NPV declined to the point of being uneconomic.

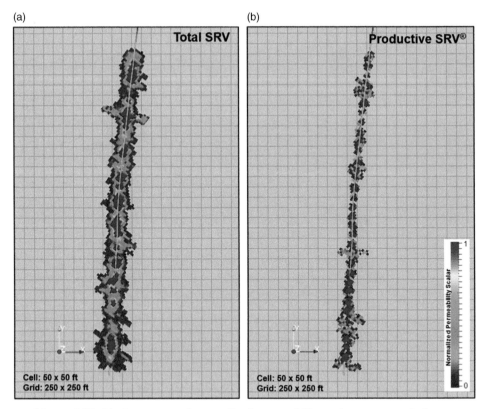

Figure 8.21 (a) A contour of normalised permeability enhancement relative to background on a horizontal slice through the wellbore for the raw DFN of Figure 8.20c. (b) The same slice as part (a) but now only the propped or productive SRV as estimated with the method described is shown. (Data courtesy of Element Petroleum)

A similar analysis is possible for stage spacing, treatment times, rates and proppant densities. In each case the value of the increased production must be weighed against the full-cycle costs of the increased development effort.

8.4.5 *Stress and Geomechanics*

So far, we have learned that the moment tensor description of the focal mechanism allows us to determine the size, the strike and the dip of the failure plane that created the microseismic event. The moment tenser description also provides us the rake or sense of motion on the plane of failure. That motion is the result of the principal stresses acting upon the failure plane at the time of failure. It is intuitively obvious that the direction of motion (rake) is driven by the ratio of the principal stresses as projected onto the failure plane (Jost and Herrman, 1989).

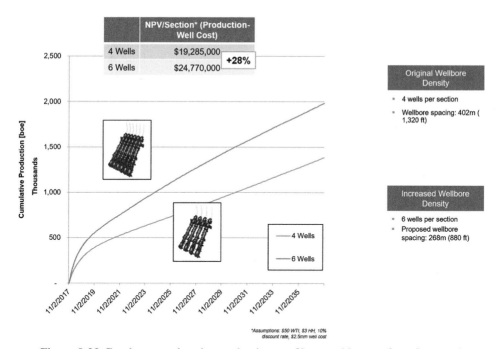

Figure 8.22 Graph comparing the production profile over 20 years for a four- and six-well development plan in the San Andres reservoir based on the observations of SRV and P-SRV created in a single well. The six-well scenario results in a 28% increase in NPV for the project and is therefore judged to be preferred over the four-well plan. (Data courtesy of Element Petroleum)

Following Agharazi (2016), we can assume a model for the stresses acting at any point decomposed into three principal components, σ_v, σ_{Hmax} and σ_{hmin}. σ_v is assumed to be vertical. The other two stresses are the maximum and minimum horizontal stresses respectively. The direction of σ_{Hmax} is estimated from the microseismic data and σ_{hmin} is assumed perpendicular to that direction. It is useful to normalise the stress tensor by the vertical stress as shown in Eq. (8.3).

$$S_{ij} = \frac{\sigma_{ij}}{\sigma_v} = \begin{bmatrix} k_H & 0 & 0 \\ 0 & k_h & 0 \\ 0 & 0 & 1 \end{bmatrix} \tag{8.3}$$

where $k_H = \sigma_{H\,max}/\sigma_v$ and $k_h = \sigma_{h\,min}/\sigma_v$

If we assume that the rake direction, as determined from the moment tensor description of the event, is parallel to the maximum shear stress acting on the slip

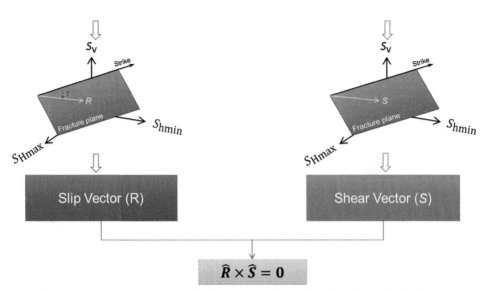

Figure 8.23 An illustration of the process for estimating the principal stress components driving the shear failure on an arbitrary fracture plane. The slip $\hat{R}$ is assumed to be in the direction of the resultant shear stress $\hat{S}$ projected onto the plane.

face as depicted in Figure 8.23, then it follows that the cross product of the slip vector with the stress tensor should be zero. In other words, it is assumed that the failure is purely shear in the direction of the resultant shear stress. Under this condition, k_H can be expressed as a linear function of k_h for each event (Agharazi, 2016).

It follows that if we can estimate σ_v from the weight of the overburden and σ_{hmin} from a diagnostic fracture injection test (DFIT) or other such test, then an estimate of σ_{Hmax} can be obtained. Such an estimate is very useful for wellbore design and reservoir modelling.

8.5 Future Developments and Other Applications of the Technology

8.5.1 Fibre-Optic Cables for Distributed Acoustic Sensing Applications

The use of fibre-optic cables as microseismic sensors was introduced in Section 8.2.5. In the energy industry fibre-optic sensing or distributed acoustic sensing (DAS) was initially utilised in the early 2000s with permanent cables attached to the outside of the casing. The fibre detected high-frequency acoustic 'noise' associated with fluid flow either out of the well for frac monitoring or into the well for production monitoring. The acoustic signals create strain in the fibre that can be measured very precisely with laser light sent down the fibre to reflect off imperfections in the fibre material (Chambers, 2019) Temperature changes at points of inflow and outflow from the well also result in strain of the fibre and can be exploited as distributed temperature sensing (DTS) (Ouyang and Belanger, 2004). However, deploying the fibre on the outside of the casing is expensive. Failures owing to breaking of the cable or darkening of the fibre as a result of chemical reactions or temperature fatiguing are reasonably common. For these reasons outside casing fibre deployment has been used on only a limited basis. Other uses of the DAS technology in this high-frequency band are for pipeline leak detection and intruder detection at borders or fences. Such applications can be accomplished with a single fibre sensor deployed at the point of need.

Low-frequency strain detection was the next application of DAS. Longer period events can be associated with what are commonly called 'frac hits', that is, when a hydraulic induced fracture actually intersects a fibre causing a strain event (Jin and Baishali, 2017). The fibre first stretches as the fracture intersects the fibre and then compresses as the pressure in the fracture leaks off and the fracture collapses. 'Frac hits' and other such frac-driven interactions (FDI) between treated wells often are associated with production issues related to well and treatment interference either at the time of treatment or over the life of the well (Daneshy and King, 2019).

It is in the mid-frequency range (3–500 Hz) that DAS becomes applicable to seismic applications in general, and microseismic mapping of hydraulic fracturing in particular, as a replacement for the geophone arrays (cf. Daley et al., 2013; Willis et al. 2016; Olofsson and Martinez, 2017). DAS data have the advantage of being broadband in time and space, since the sampling interval down the fibre can be on the order of centimetres (Harthog, 2018). However, the DAS signal is only a measure of strain along the direction of the cable, a one-dimensional observation. If one considers how event location is achieved as described earlier, it is clear that observations from a single fibre sensor are severely constrained as far as event location is concerned. Essentially such an observation set can only indicate distance from the cable but not direction. If the fibre is deployed in both the vertical and horizontal portions of the well, then an improved location estimate is available, at least for those events close enough to the vertical portion of the well to be registered. A better solution is to have three monitor fibres in separate wells, all within detection distance. Temporary or intervention style fibre tools which are placed into existing wells only for the duration of monitoring (Attia et al., 2019) make this economically feasible when the appropriately spaced wellbores are available.

8.5.2 Other Applications of Microseismic Monitoring

Microseismic monitoring has many applications in the energy and resource industries other than hydraulic fracture well stimulation monitoring. There is a long history of application in mining for the early detection of rock bursts and caving (e.g., Gibowicz and Kijko, 1994). An early application of microseismic monitoring was used to map the collapse structure at Ekofisk in the Norwegian North Sea as a result of fluid injection/production (Maxwell et al., 1996, 2001). The delineation of faulting showed compartmentalisation of the reservoir and demonstrated microseismic monitoring as a reservoir management tool. Another example of microseismic monitoring applied to a North Sea chalk reservoir is discussed in the Valhall case study in Section 8.6.2. Microseismic monitoring has also been applied in the development of heavy oil resources through steam injection (e.g., Maron et al., 2005; Maxwell et al., 2008) and has been suggested as a tool for long term production monitoring (e.g., Jupe et al., 2000).

Monitoring of the frac'ing to develop geothermal resources regularly use techniques similar to those described here (cf. Liaw et al., 1979; Dyer et al., 2008). Monitoring has been applied to CO_2 sequestration projects (Maxwell et al., 2004) to track fluid movement. The recent interest in induced seismicity related to frac'ing and wastewater disposal has led to the monitoring of injection wells for mitigation of seismic hazards (Zoback, 2012; Eaton, 2018). Finally, there has been a limited number of studies attempting to use both induced and naturally occurring microseismic events or other ambient noise as seismic sources for exploration seismic imaging by either reflection, tomography or interferometry (Artman, 2006; de Ridder and Dellinger, 2011).

8.6 Case Studies

8.6.1 Hydraulic Stimulation for Tight Oil Production: Vaca Muerta Shale, Argentina

The Vaca Muerta shale oil play in Argentina's Neuquén Basin (Figure 8.24) is estimated to contain technically recoverable reserves of 308 trillion cubic feet of gas and 16 bbls of oil. It is likely to make a large contribution to Argentina's hydrocarbon production in the future. Wintershall Energia acquired operatorship of a part of this play in 2014 (Figure 8.25) and decided to apply microseismic monitoring to its well completion programmes. Reservoir management required frac'ing of the Vaca Muerta formation to enhance permeability (by inducing new or reactivating natural fractures) and hence improve well recovery efficiency.

The reservoir development plan included both vertical and horizontal wells. The goal of the monitoring was to inform decisions on horizontal wellbore direction,

Figure 8.24 Hydrocarbon producing basins of Argentina with the location of the Neuquén Basin noted. (After Curia et al., 2018)

wellbore and stage spacing and landing depth. The technical strategy was to develop the monitoring programme in two stages:

1. An initial test using a surface array to inform design of a permanent buried array to be used for monitoring a large number of future development wells
2. Installation of a buried array spread over an area of 5 × 7 km for permanent monitoring of development wells

Stage 1 of the project (surface monitoring) was successful in detecting events created from treating the first development well (AF.x-1; see Curia et al., 2018). Analyses of these records in the context of existing well log and seismic data

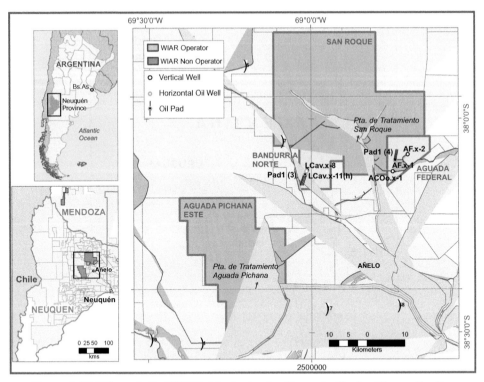

Figure 8.25 Wintershall (WIAR) blocks in the Neuquén Basin. The wells discussed here are in the Aguada Federal block. (After Curia et al., 2018)

showed that most events correlated with the Upper Vaca Muerta formation and few from the Lower unit. This resulted in a recommendation to land development wellbores in the Upper Vaca Muerta formation which was an important management decision. It also provided guidance on the optimal orientation of future development wellbores.

This initial phase also enabled detailed design for a permanent near surface buried array. The important design considerations for the establishment of such an array are as follows:

- Depth of geophone placement (to achieve optimal signal/noise vs cost of drilling)
- Individual station array configuration
- Areal extent of the array (aperture)
- Number of stations (fold)

The second monitoring (pilot phase) utilised a buried array based on the earlier work. Four development wells drilled between December 2016 and May 2017 were monitored as they were completed. Each of these wells was landed in the Middle Vaca Muerta (Curia et al., 2018).

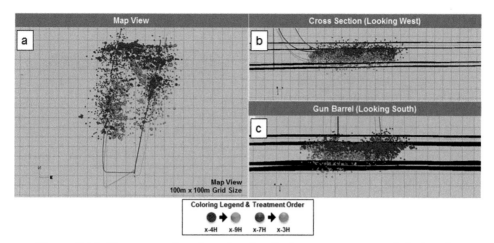

Figure 8.26. Hypocentre locations for the final event set captured during the treatment of the four horizontal wells. (a) Map view. (b) Depth view looking west. (c) Gun barrel view looking south. Horizon tops are shown for Upper, Middle and Lower Vaca Muerta as well as the underlying Tordillo. Grid size in all plots is 100 m × 100 m. Events are sized by magnitude (Curia et al., 2018).

The distribution of the seismic hypocentres following treatment of these wells is shown in Figure 8.26.

A discrete fracture network (DFN) was constructed from the microseismic event catalogue. This gives a geologically more reasonable model that can be upscaled for use in a reservoir simulator (Curia et al., 2018). Figure 8.27 displays a map and depth section view of the DFN derived from the microseismic monitoring of the four horizontal wells that were part of the Pilot Phase.

This monitoring example, the first such successful monitoring of frac'ing in the Vaca Muerta, was part of the implementation of a typical shale development program having four phases: technical, pilot, pre-development and development. This case describes the monitoring of microseismic events created by frac'ing treatments in the first two phases: technical, to prove hydrocarbon flow, and pilot, to reduce uncertainty in production rates.

During the technical phase, the monitoring of a single vertical well with a large surface array established that surface-based sensors could indeed detect events originating in the Vaca Muerta.

The pilot phase observations provided a detailed description of the fracture network created by the treatments. Focal mechanisms determined for the detected events were used to understand the stress distribution in the reservoir and to further refine the completion (wellbore orientation, spacing and stage spacing) parameters to be used going forward.

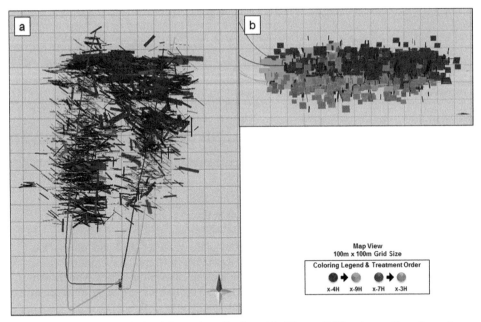

Figure 8.27. DFN derived from data displayed in Figure 8.26 using strike, dip and magnitude of the observed microseismic events. (a) Map view. (b) depth view looking west. Grid size is 100 m × 100 m (Curia et al., 2018).

8.6.2 Microseismic Monitoring of the Valhall Chalk Reservoir, North Sea

North Sea chalk reservoirs such as Valhall in the Norwegian Offshore Continental Shelf (NOCS) are well known for their sea floor subsidence associated with reservoir compaction and chalk collapse. Valhall is a giant oil field in Blocks 2/8 and 2/11 in the Norwegian sector of the Central Graben of the North Sea (Figure 8.28). It was discovered in 1975 and production commenced in 1982 from the Late Cretaceous Tor chalk reservoir at a depth of around 2,400 m with water depths being around 70 m. The total in place volume was estimated at 2.74 Bboe with initial reserves of only 250 mmboe. Twenty years later the total production had reached 625 mmboe and the field was producing on plateau at 60,000 boepd with economic production expected to continue until 2050 (Barkved et al., 2003).

The field has, however, proved to be difficult to manage. The high porosity, low permeability chalk reservoir rock is weak and collapses under production. This compaction results in subsidence over the reservoir that affects the entire overburden up to the seafloor. This has complicated the drilling of many of the wells needed to develop the field and required the central facilities to be jacked up.

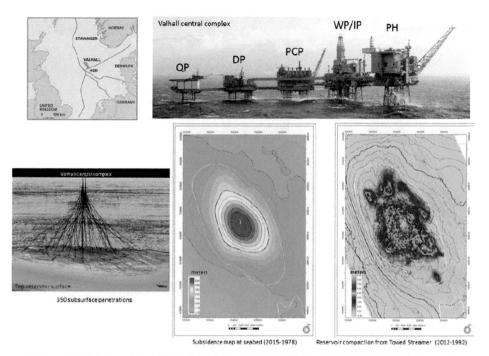

Figure 8.28 Map of the Valhall Field in the NOCS (top left), the Valhall Central Complex (top right), the (2015) Seabed Subsidence map (bottom middle) and the 2012 Reservoir Compaction Map (bottom right) (Haller et al., 2016).

Although the average rate of subsidence reduced from 25 cm/year in the first 20 years of production to 11 cm/year over the next decade the central facilities were moved and re-developed in 2013 so as to be able to extend the life of the field. Maps of the reservoir compaction up to 2012 and the seabed subsidence up to 2015 are shown in Figure 8.28 (Haller et al., 2016).

The chalk reservoir is generally complex and there are gas-charged sediments in the overburden. The presence of the associated gas cloud over the crest of the structure makes it very difficult to image the reservoir with P-wave data from conventional towed streamer seismic. Shear wave data are relatively unaffected by the gas in the overlying 'gas cloud'. Using ocean bottom recording to detect converted PS waves enabled much better imaging of the reservoir and the faults associated with the subsidence (Barkved et al., 2004).

In 1998 a microseismic monitoring study was carried out to improve the imaging of the fault structure. A 100 m long wireline array consisting of six three-component geophones spaced at 20-m intervals above the reservoir was deployed for 57 days in a well near the crest of the field to monitor hydraulic fracture treatments. Results of the survey were reported by Barkved et al. (1999). The

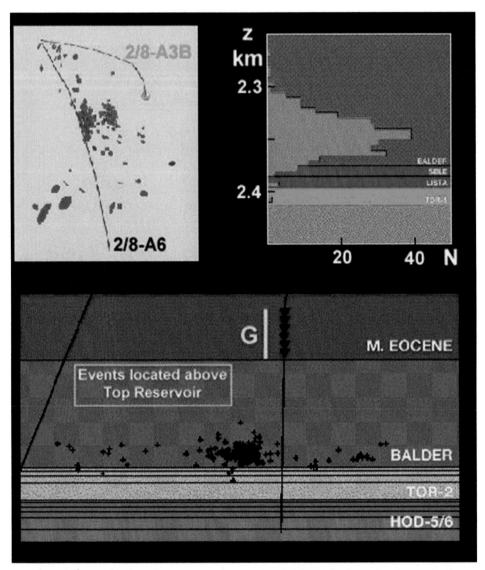

Figure 8.29 Location of events detected during 57 days of microseismic monitoring at Valhall (Barkved et al., 1999).

location of the events detected are displayed in Figure 8.29. Around 10 events per day were recorded and 324 of the 572 events detected were initially located. Nearly all the events occurred within a 50-m zone above the top of the Balder Formation reservoir (the base of the grey shading in the depth sections in Figure 8.29) where wellbore stability problems were common (Kristiansen et al., 2000).

The data were later analysed in more detail at the University of Leeds (De Meersman et al., 2009). More than 300 of the events were re-located and the location uncertainty reduced to half that of the previous studies. This clearly demonstrates the way in which location accuracy can be improved by careful processing, although the time involved in such painstaking work can be considerable so the operational value of this improvement may be limited.

Further work on the borehole data set by De Meersman et al. (2009) also reported the results of shear-wave splitting analysis which, integrated with the relocated source positions, support a model whereby stresses at Valhall recharge cyclically. Stress builds up in response to reservoir compaction and is released by microseismic activity. These changes result in temporal variations in seismic anisotropy and focal mechanisms.

In 2003 BP installed a dense array of four component seismic sensors covering an area of 45 km^2 on the seabed above the Valhall field (Kommedahl et al., 2004). This 'life of field seismic' (LOFS) array was the first full-scale permanent reservoir monitoring (PRM) LOFS installation intended to allow cost-effective, regular, repeat 3D seismic surveys for reservoir monitoring over the remaining life of the field. The cost of the installation (including the 120 km of cables and the trenching was around $40 million with the acquisition cost of each repeat 4D survey being of the order of $500k. A total of more than 20 4D surveys were subsequently undertaken at intervals of 6–12 months and the value of the initial installation repaid many times over in terms of the better well planning and increased production.

Although not designed for microseismic applications, the LOFS array also allowed for microseismic monitoring to be carried out and Chambers et al. (2010) demonstrated that such an array can be used to monitor events induced by hydraulic fracturing. Vertical component data from 799 stations nearest to one of the injection wells were analysed and over 1,000 candidate events were identified during one 6.5-hour fracturing operation. The events extended approximately 300 m vertically and 200 m horizontally with sources both within the chalk layer containing the reservoir and the overburden. The temporal distribution of the events was found to be inconsistent with the induced fracturing extending into the overburden. Instead, the synchronous increase in event activity with the application of the downhole pressure was interpreted as evidence that elastic wave fronts originating from the well activity triggered movement on faults where the in situ stress is close to that required for failure. The largest cluster of events appeared to define a steeply dipping zone of faults that were not aligned with the overall structure. Knowing the location of these faults had an impact on the subsequent production planning.

References

Agharazi, A., 2016. Determination of maximum horizontal field stress from microseismic focal mechanisms: A deterministic approach. In *50th U.S. Rock Mechanics/ Geomechanics Symposium*, 26–29 June 2016, Houston, TX.

Aki K. and Richards, P. G., 2002. *Quantitative Seismology*, 2nd ed. Herndon, VA: University Science Books.

Artman, B., 2006. Imaging passive seismic data. *Geophysics* **71**(4), SI177–SI187. DOI: 10.1190/1.2209748.

Attia, A., Brady, J., Lawrence, M. and Porter, R., 2019. Validating refrac effectiveness with carbon rod conveyed distributed fiber optics in the Barnett Shale for Devon Energy. *Society of Petroleum Engineers. SPE* 194338. DOI: 10.2118/194338-MS.

Baig, A. and Urbancic, T., 2010. Magnitude determination, event detectability and assessing the effectiveness of microseismic monitoring programs in petroleum applications. *CSEG Recorder*, 22–26 February.

Baig, A, Urbancic, T. and Seibel, M., 2010. The effect of microseismic array configuration on the determination of hydraulic fracture parameters. *CSEG Annual Meeting and Exhibition.*

Barkved, O., Dyer, B. C., Jones, R. H. and Folstad, P. G., 1999. Microseismic monitoring of the Valhall Field. *Extended Abstracts, EAGE 61st Conference*, Helsinki.

Barkved, O. I., Heavey, P., Kyelstadli, R., Kleppen, T. and Kristiansen, T. G., 2003. Valhall Field: Still on plateau after 20 years of production. In *Proceedings SPE European Conference*, Paper 83957.

Barkved, O., Kommedahl, J. H. and Thomsen, L., 2004. The role of multi-component seismic data in developing the Valhall field, Norway. *Extended Abstracts E040, EAGE 61st Conference*, Helsinki.

Brune, J., 1970. Tectonic stress and spectra of shear waves from earthquakes. *Journal of Geophysical Research*, **75**, 4997–5009.

Bulant, P., Eisner, L., Psencik, I. and Le Calvez, J., 2007. Importance of borehole deviation surveys for monitoring of hydraulic fracture treatments. *Geophysical Prospecting*, **55**, 891–9. DOI: 10.111/j.1365-2478.2007.00654.x.

Cao, R., Li, R., Girardi, A., Chowdhury, N. and Chen, C., 2017. Well interference and optimum well spacing for Wolfcamp development at Permian Basin. *Unconventional Technology Conference (URTeC).* DOI: 10.15530/urtec-2017-2691962.

Carlson, T. C. and Sipes, L. D. Jr., 1965. Characteristics of a San Andres reservoir. SPE-1145-MS. DOI: 10.2118/1145-MS.

Chambers, K., 2019. What is DAS and what is it measuring? https://motionsignaltechnologies.com

Chambers, K., Kendall, J-M. and Barkved, O., 2010. Investigation of induced seismicity at Valhall using the Life of Field seismic array. The Leading Edge, **29**(3), 290–5.

Cieslewicz, D. and Lawton, D. C., 1996. Receiver notching in a linear $V(z)$ near-surface medium. CREWES Research Report 10, chapter 3.

Crovetto, C., Moirano, J., Vernengo, L., et al., 2020. Imaging a two-lateral zipper frac with a surface microseismic array in Vaca Muerta, Argentina. In *Unconventional Resources Technology Conference (URTeC).* DOI: 10.15530/urtec-2020-1503.

Curia, D, Duncan, P. M., Grealy, M., McKenna, J. and Hill, A., 2018. Microseismic monitoring of Vaca Muerta completions in the Neuquén Basin, Argentina. *The Leading Edge*, **37**(4), 262–9. DOI: 10.1190/tle37040262.1.

Daley, T. M., Freifeld, B. M., Ajo-Franklin, J., et al., 2013. Field testing of fiber-optic distributed acoustic sensing (DAS) for subsurface seismic monitoring. *The Leading Edge*, **32**(6). DOI: 10.1190/tle32060699.1.

Daneshy, A. and King, G., 2019. Frac-driven interaction (FDI) between horizontal wells: Causes, consequences and mitigation techniques. *Hydraulic Fracturing Journal*, **5**(4), 4–28.

de Ridder, S. and Dellinger, J., 2011. Ambient seismic noise eikonal tomography for near surface imaging at Valhall. The Leading Edge, **30**(5), 506–12. DOI: 10.1190/1.3589108.

Duncan, P. and Eisner, L., 2010. Reservoir characterization using surface microseismic monitoring. *Geophysics*, **75**, 75A139–75A146.

Duncan, P. M., Smith, P. G., Smith, K. W., Barker, B., Williams-Stroud, S. and Eisner, L., 2013. Microseismic monitoring in early Haynesville development. In U. Hammes and J. Gale (eds.), *Geology of the Haynesville Gas Shale in East Texas and Louisiana, USA*. AAPG Memoir **105**, 217–34.

Dyer, B. C., Schanz, U., Ladner, F., Häring, M. O. and Spillman, T., 2008. Microseismic imaging of a geothermal reservoir stimulation. *The Leading Edge*, **27**(7), 856–69. DOI: 10.1190/1.2954024.

Eaton, D. W., 2018. *Passive Seismic Monitoring of Induced Seismicity*. Cambridge: Cambridge University Press.

Eaton, D. W., and Forouhideh, F., 2011. Solid angles and the impact of receiver array geometry on microseismic moment-tensor inversion. *Geophysics*, **76**(6), WC77–WC85.

Eisner, L., Heigl, W., Duncan, P. M. and Keller, W., 2009. Uncertainties in passive seismic monitoring. *The Leading Edge*, **28**(6), 648–55.

Eisner, L., Williams-Stroud, S., Hill, A., Duncan, P. and Thornton, M., 2010. Beyond the dots in the box: Microseismic-constrained fracture models for reservoir simulation. *The Leading Edge*, **29**(3), 326–33. DOI: 10.1190/1.3353730.

Evans, D. M., 1966. The Denver area earthquakes and the Rocky Mountain Arsenal disposal well. *The Mountain Geologist*, **3**(1), 23–36.

Gibowicz, S. J. and Kijko, A., 1994. *An Introduction to Mining Seismology*. San Diego: Academic Press.

Grechka, V. and Heigl, W. M., 2017. *Microseismic Monitoring*. Geophysical References Series No. 22. Tulsa, OK: Society of Exploration Geophysicists.

Haller, N., Flateboe, R., Twallin, C., et al., 2016. Valhall case study: Value of seismic technology for reducing risks in a reactive overburden. *Extended Abstracts, 78th EAGE Conference and Exhibition*, Vienna, Austria.

Harthog, A. H., 2018. *An Introduction to Distributed Optical Fibre Sensors*. Boca Raton, FL: CRC Press.

Jin, G. and Baishali, R., 2017. Hydraulic-fracture geometry characterization using low-frequency DAS signal, *The Leading Edge*, **36**(12), 975–80. DOI: 10.1190/tle36120975.1.

Jost, M. L. and Herrmann, R. B., 1989. A student's guide to and review of moment tensors. *Seismological Research Letters*, **60**(2), 37–57.

Jupe, A., Jones, R., Wilson, S. and Cowles, J., 2000. The role of microearthquake monitoring in hydrocarbon reservoir management. In *SPE Annual Technical Conference and Exhibition*, 1–4 October 2000, Dallas, Texas. DOI: 10.2118/63131-MS.

Kashikar, S., Shojaei, H. and Lipp, C., 2015. Accurate modelling improves early production predictions. *World Oil*, November, 16–19.

Khodabakshnejad, A., Rahimi Zeynal, A. and Fontenot, A., 2019. The sensitivity of well performance to well spacing and configuration: A Marcellus case study. In *Unconventional Resources Technology Conference (URTeC)*. DOI: 10.15530/urtec-2019-1076.

Kommedahl, J.H., Barkved, O. I. and Howe, D. J., 2004. Initial experience operating a permanent 4C seabed array for reservoir monitoring at Valhall. *Extended Abstract, SEG 74th Annual Meeting*, **23**, 2239–42.

Kristiansen, T., Barkved, O. and Patillo, P., 2000. Use of passive seismic monitoring in well and casing design in the compacting and subsiding Valhall field. In *Proceedings SPE European Conference*, Paper 65134.

Lay, T. and Wallace, T. C., 1995. *Modern Global Seismology*, San Diego: Academic Press.

Liaw, A. L. and McEvilly, T. V., 1979. Microseisms in geothermal exploration: Studies in Grass Valley, Nevada. *Geophysics*, 44(6), 1097–115.

Maron, K. P., Bourne, S., Wit, K. and McGillivray, P., 2005. Integrated reservoir surveillance of a heavy oil field in Peace River, Canada. In *EAGE 67th Conference and Exhibition*, C034.

Mayerhofer, M. J., Lolon, E. P., Warpinski, N. R., Cipolla, C. L., Walser, D. and Rightmire, C. M., 2010. What is stimulated reservoir volume? *SPE Production & Operation*, 25(1), 89–98, 119890.

Maxwell, S., 2014. Microseismic imaging of hydraulic fracturing: Improved engineering of unconventional shale reservoirs. Distinguished Instructor Series, No. 17, Society of Exploration Geophysicists.

Maxwell, S. C., Du, J. and Shemeta, J., 2008. Passive seismic and surface monitoring of geomechanical deformation associated with steam injection. *The Leading Edge*, **27** (9), 1176–184. DOI: 10.1190/1.2978980.

Maxwell, S. C., Rutledge, J., Jones, R. and Fehler, M., 2010. Petroleum reservoir characterization using downhole microseismic monitoring. *Geophysics*, **75**, 75A129–75A137.

Maxwell, S. C. and Urbancic, T. I., 2001. The role of passive microseismic monitoring in the instrumented oil field. *The Leading Edge*, **20**(6), 636–9.

Maxwell, S. C., Urbancic, T., Steinsberger, N. and Zinno, R., 2002. Microseismic imaging of fracture complexity in the Barnett Shale. In *Proceedings of the 2002 Societyof Petroleum Engineers, Annual Technical Conference and Exhibition*, San Antonio, TX, paper 77440.

Maxwell, S. C., White, D. J. and Fabriol, H., 2004. Passive seismic imaging of CO_2 sequestration at Weyburn. *Expanded Abstracts, SEG Technical Program*. DOI: 10 .1190/1.1842409.

Maxwell, S. C., Young, R. P., Bossu, R., Jupe, A. and Dangerfield, J., 1996. Microseismic logging of the Ekofisk Reservoir. Paper presented at the SPE/ISRM Rock Mechanics in Petroleum Engineering, Trondheim, Norway, July 1998. DOI: 10.2118/47276-MS.

Neuhaus, C. W., McKenna, J., Rahimi Zeynal, A., Telker, C. and Ellison, M., 2014. Completions and reservoir engineering applications of microseismic data. *Expanded Abstracts, SEG Technical Program*, 4564–9. DOI: 10.1190/segam2014-1365.1.

Oda, M., 1985. Permeability tensor for discontinuous rock masses. *Geotechnique*, **35**(4), 483–95.

Olofsson, B. and Martinez, A., 2017. Validation of DAS data integrity against standard geophones: DAS field test at Acquistore site. *The Leading Edge*, **36**(12), 981–6. DOI: 10.1190/tle36120981.1.

Ouyang, L. and Belanger, D., 2004. Flow profiling via Distributed Temperature Sensor (DTS) system: Expectation and reality. In *SPE Annual Technical Conference and Exhibition.*

Raleigh, C. B., Healy, J. H. and Bredhoeft, J. D., 1976. An experiment in earthquake control at Rangely, Colorado. *Science*, New Series, **191**(4233), 1230–7.

Rutledge, J. T., Phillips, W. S., House, L. S. and Zinno, R. J., 1998. Microseismic mapping of a Cotton Valley hydraulic fracture using decimated downhole arrays. *Expanded Abstracts, Society of Exploration Geophysicists 68th Annual International Meeting*, 338–41.

Shapiro S. A., Rothert, E., Rath, V. and Rindschwentner, J., 2002. Characterization of fluid transport properties of reservoirs using induced microseismicity. *Geophysics*, **67**, 212–20.

Vavryčuk, V., 2007. On the retrieval of moment tensors from borehole data. *Geophysical Prospecting*, **55**, 381–91.

Wessels, S. A., De La Peña, A., Kratz, M., Williams-Stroud, S. and Jbeili, T., 2011. Identifying faults and fractures in unconventional reservoirs through microseismic monitoring. *First Break*, 29(7). DOI: 10.3997/1365-2397.29.7.51919.

Williams-Stroud S., 2008. Using microseismic events to constrain fracture network models and implications for generating fracture flow properties for reservoir simulation. Paper presented at the SPE Shale Gas Production Conference, November 2008, Fort Worth, TX. DOI: 10.2118/119895-MS.

Williams-Stroud, S. and Eisner, L.., 2009. Method for determining discrete fracture networks from passive seismic signals and its application to subsurface reservoir simulation. Patent no. US8902710B2.

Williams-Stroud, S., Kilpatrick, J. E., Cornette, B., Eisner, L. and Hall, M., 2010. Moving outside the borehole: Characterizing natural fractures through microseismic monitoring. *First Break*, **28**(7), 89–94.

Willis, M. E., Barfoot, D., Elmauthaler, A., et al., 2016. Quantitative quality of distributed acoustic sensing vertical seismic profile data. *The Leading Edge*, **35**(7). DOI: 10.1190/tle35070605.1.

Xiong, H., Ramanathan, R. and Nguyen, K., 2019. Maximizing asset value by full field development: Case studies in the Permian Basin. In *Unconventional Technologies Conference (URTeC)*. DOI: 10.15530/urtec-2019-554.

Zoback, M. D., 2012. Managing the seismic risk posed by wastewater disposal. *Earth Magazine*, April, 38–43.

9

A Road Map for Subsurface De-risking

HAMISH WILSON

The preceding chapters describe each technology on a standalone basis and explain its application and integration potential. This section pulls the whole story together and outlines the thinking process and approach to selecting the optimal technology and integration approach.

A typical exploration opportunity is evaluated using a combination of 2D and 3D seismic followed up with drilling a well. Chapter 2 describes the industry workflows and evaluation processes and how these are generally optimised around these core technologies which are most heavily applied during exploration and development in mature and well-known provinces.

This book demonstrates that there are other approaches that when used alone, or in conjunction with seismic, can considerably reduce the uncertainty and risk of a given opportunity. Non-seismic technologies have a risk reduction role throughout the exploration value chain, from frontier basin entry exploration through to development and production and on into carbon capture and storage.

Integration is a theme running through our commentary alongside designing the survey with an understanding of the geological model you plan to validate. This is particularly the case with gravity/magnetics, Full Tensor Gradiometry (FTG) and electromagnetic (EM) techniques for which there are non-unique outcomes for a given response. Hence their best application is when used in combination with either refraction seismic or conventional reflection seismic.

As well as providing direct imaging in their own right, gravity, EM methods and refraction seismic add significant value in helping to greatly improve the migration of seismic data. This is particularly the case when seismic data quality is compromised due to the presence of difficult lithologies in the overburden (e.g., salt, basalt and chalk) or to increase the resolution at depth.

We present a very timely review of marine EM methods, in which we explain that if correctly applied, in the right geological conditions, marine EM has an important role to play in prospect de-risking and reservoir management.

9.1 Where in the Exploration Value Chain?

Table 9.1 summarises in which stage of the exploration value chain the specific technology should be used.

Questions that need addressing early in the exploration cycle include

- Is there a basin?
- Is there a working petroleum system?
- What is the overall structure of the basin?

Gravity and magnetic surveys are ideal in providing the answers cheaply and effectively. Magnetotelluric methods can also provide valuable information in assessing these questions.

Basin structure, thermal history and tectonic mechanisms are important building blocks for understanding the petroleum system. Understanding the crustal structure goes a long way to providing clues prior to further data acquisition, and in some circumstances use of deep crustal geophysics is justified. Thus, as one progresses further into the exploration opportunity, there is a requirement to increase the granularity of the data analysis to understand the scale of the target.

Basin evaluation these days is increasingly based around the use of play based methods. Play-based exploration requires holistic understanding of basin evolution and analysis of tectonic history, distribution of mega-sequences and the development of detailed sequence stratigraphic based gross depositional environment maps to be able to identify prospective areas.

Table 9.1 reflects this overall approach to exploitation. Starting from the left the need is for understanding the overall crustal and regional structural framework. This is usually best served by screening geophysics tools such as deep crustal seismic, potential fields and regional EM. To deepen our understanding of a given play, (usually at the pre-access stage) more detailed analysis is needed, which requires higher resolution techniques (2D seismic, FTG, CSEM). The greatest level of detail is needed, of course, at the prospect definition and development stages.

It is very important to note that no one method stands alone at any stage of the value chain. Integration of various data sources to test for consistency is critical. It is this that helps constrain uncertainty and risk envelopes. Reliance on any one method on its own has frequently led to serious and expensive mistakes, often because of some pitfall that could have been avoided if other information had been integrated.

9.2 Geoscience Issues Analysis

Another approach to identifying which technologies might help in any given situation is to take a more functional view. Here we ask what each technology can

Table 9.1 *Geophysical technologies and the exploration value chain*

E&P Phase	Basin screening	Access	Exploration	Prospect evaluation	Appraisal	Development	Production
Crustal geophysics	●						
Gravity and magnetics	●	●	○				
Full tensor gravity		●	●	●	●		
Marine CSEM and MT	●	●	●	●	●	○	○
Ocean bottom seismic			●	●	●	●	●
Microseismic and passive seismic						●	●
Conventional reflection seismic	○	●	●	●	●	●	●

CSEM, controlled-source electromagnetic; MT, magnetotelluric.
Green indicates a good fit; orange, partial fit; blank, not applicable.

deliver in terms of impact on a geophysical or geological question. The individual chapters in this book have identified specific geoscience problems and issues that the given technology can address. These issues have been collated into a single summary table (Table 9.2) that represents the complete range of problems that these technologies can address and gives an indication of whether the given method should be considered to help address a particular geoscience problem.

Against each geoscience issue we have tabulated the technologies that can address that problem. We have used the traditional traffic light system to highlight effectiveness as follows:

Green indicates that you should always consider using that technology.

Yellow indicates that the technology provides a partial solution and might sometimes be effective. For the most part, the issue is one of resolution and being able to discern the detail of features (see also Table 9.2).

Red recommends that the technology should not be used.

Grey indicates that the technology is not applicable.

At the top of each technology column we have added an indicative cost of a typical survey based on a very broad average over the last five to ten years, either as a per km or km^2 cost, or for a typical deployment. We have also added, for reference, columns for 2D and 3D seismic on- and offshore and their respective order of magnitude costs. It is clear from a simple cost comparison that especially for onshore exploration, non-seismic technologies can add very significant value.

The table categorises the geoscience problems into logical groupings that approximately correspond to the exploration value chain: Basin, Play Fairway, Prospect and Reservoir.

9.2.1 Basin

Basin formation mechanism (and thus heat flow history) and crustal thickness can be best determined using crustal seismic techniques in conjunction with regional potential field methods. The other technologies are either not applicable or should not be used (mainly because of too high a cost at this stage of the exploration process).

Gravity, magnetics and MT data are very powerful tools for understanding basin architecture. These techniques are valuable for determining depth to basement and basin configuration and are cost effective. Notice that the application of seismic at the mega-regional scale is conditioned mainly by cost. Certainly, 3D in early-stage exploration, or when the basin architecture is poorly understood, is not generally cost effective. Long-offset 2D is clearly of value and is rightly being increasingly used even for basin architecture studies given the importance of understanding

Table 9.2 *Geophysical technologies and the geoscience problem*

Geoscience problem	Crustal seismic	AeroMagnetics	Gravity	FTG	MT	CSEM	OBN	Microseismic[b]	Land 2D seismic	Marine 2D seismic	Land 3D seismic	Marine 3D seismic
Cost $/stn deployment or frac stage monitored[a]	12,000–16,000							5,000–20,000				
Cost $/km (based on average costs 2014-2020)	3,000–4,000	15–20	100 (ground) 80 (air)	150	Survey dependent	Survey dependent		50,000–300,000	4,000–20,000	1,000–2,000		
Cost $/km² (based on average costs 2014-2020)							30,000–200,000	5,000–25,000			15,000–100,000	8,000–30,000
Basin												
Tectonic basin formation mechanism												
Crustal thickness								Earthquakes				
Heat flow history												
Basin delineation												
Depth to basement								Earthquake/ noise		Long offset		
Overthrust configuration								Earthquakes				
Basin fabric												
Major fault delineation								If faults active				
Play Fairway Analysis												
Fault delineation								If faults active				
Cross section modelling												
Detection/ delineation of volcanics												
Velocity constraints/ modelling												
PSDM modelling												
Seismic migration improvement												
Imaging prospect scale												

322

Prospect	Clastic/salt interface
	Clastic carbonate interface
	Prospect definition
	PSDM modelling
	Prospect definition at relatively shallow depth (~<2 km)
	Sub-basalt imaging
	Subsalt imaging
	Lithology prediction
	Fluid prediction
Production and Reservoir Management	Production monitoring
	PSDM modelling
	Location of old wells and pipelines
	Fault location
	Fault mobility
	Fracture orientation
	Fracture location

LEGEND

Always Consider	Partial solution- sometimes effective	Not appropriate at this stage - not cost effective	Not Appropriate

[a] Costs are for proprietary surveys (not multi-client).

[b] Costs include acquisition and processing.

CSEM, controlled-source electromagnetic; FTG, full tensor gradiometry; MT, magnetotelluric; OBN, ocean bottom node.

Green indicates a good fit; orange, partial fit; red indicates not appropriate at this stage, and not cost effective; blank, not applicable.

tectonic history. Notice the 'niche' contribution that microseismic can make to locate active faults; potentially of real value for onshore unconventional hydrocarbon production.

To conclude, an ideal basin architecture work flow should therefore integrate crustal seismic, potential field data alongside long-offset 2D for the best results.

9.2.2 Play Fairway Analysis

The specific problems that these technologies can address can be grouped into two sets, firstly, lithology determination (volcanics, salt, carbonates) and secondly providing velocity data for seismic imaging. Both FTG and CSEM have important roles in supporting 2D and 3D seismic in reducing uncertainty in play evaluation.

Airborne magnetics and gravity surveys can add value but may not have sufficient resolution to determine play or prospect scale features. However, given a large unexplored onshore basin, airborne potential field methods can identify prospective areas in a basin that can then be used to position 2D and 3D surveys and therefore lead to a more cost-effective deployment of conventional seismic.

9.2.3 Prospect Identification

Ocean bottom nodes (OBNs) are the emerging technology that could have the biggest impact on offshore prospect identification. The key barrier to their widespread deployment is cost. OBNs are currently more than twice as expensive as traditional towed streamer 3D techniques. However, further cost reductions are expected, and these will make OBNs more competitive with conventional 3D. Clearly for a given expected risk reduction, then OBNs can be justified on the basis of a value of information analysis.

FTG and CSEM/MT have contributions to make, in particular in assisting imaging prospects at relatively shallow depths (less than 3,000 m). Both these techniques are capable of producing prospect scale definition.

Historically CSEM has been mainly used to characterise the fluid content of prospects. Experience has shown that this application, while viable, needs to be undertaken with great care. Provided CSEM is integrated into the exploration work flow, alongside other methods, it can prove effective. Thus given a seismically defined structure and information on lithology from well log data, CSEM can provide information to de-risk the likely fluid content. Seismic data provide information on the reservoir structure (but usually not fluid content or extent), and seismic and well log information provide information on the surrounding strata in which the reservoir is embedded. The CSEM interpretation problem is therefore better constrained, and the resulting interpretation risk is consequently lower.

Statistics quoted by Shell suggest a discovery success rate of 80% through the integration of CSEM surveys into the exploration workflow (compared to approximately 47% before the integration of CSEM), with a 100% success in the prediction of dry holes pre-drill (Karman et al., 2011).

9.2.4 Production and Reservoir Management

Currently 3D seismic acquired using OBN technology offshore gives the best reservoir images. The technology can provide a remarkable imaging uplift over conventional marine 3D, albeit at a significantly higher cost. This cost increase can often easily be justified on the basis of enhanced reservoir description.

Micoseismic is proving to be an essential tool to identify and monitor faults and fractures. Fracking by its nature fractures rocks and can also remobilise existing faults; both of these effects can be detected by using microseismic techniques.

Onshore and offshore field decommissioning is becoming an important concern. and in this context, aeromagnetics has a valuable application in locating old wells and pipelines.

9.2.5 Geoscience Issues: Scale

Note that a number of the technologies provide partial solutions to the given problem. An alternative way of reviewing their application is to consider scale. Table 9.3 shows the technologies that can be considered when analysing problems at various geographical scales. The usefulness or need for special techniques often depends on whether there is an existing conventional seismic database; hence Table 9.3 also indicates the type of seismic that is likely to be available at that stage of the exploration process. As discussed earlier, these technologies are best used together and in conjunction with seismic.

The table lists the one or two technologies, that if used alone (in the case of crustal studies) or in combination with 2D or 3D seismic offers the best solution to determining that particular geoscience issue.

9.2.6 The Geological View

Figures 9.1–9.3 give a geological view of when to consider individual or combinations of these special technologies by considering common tectonic settings.

Table 9.3 Geophysical technologies and the question of scale

		Seismic coverage assumption						
		No 2D	2D	2D	2D/3D	3D	3D	3D
Geoscience problem		Basin scale	Tectonic element scale	Feature scale	Petroleum system	Play definition	Prospect scale	Reservoir scale
Basin	Tectonic basin formation mechanism	Crustal seismic	Crustal seismic					
	Crustal thickness	Crustal seismic	Crustal seismic		Crustal seismic			
	Heat flow history	Crustal seismic			Crustal seismic			
	Basin delineation	Grav/Mag/MT	Grav/Mag/MT		Crustal seismic			
	Depth to basement	Crustal seismic/MT/Mag MT			Crustal seismic			
	Overthurst configuration							
	Basin fabric	Grav/Mag/MT	Grav/Mag/MT	Grav/Mag/MT	Grav/Mag/MT			
	Major fault delineation	Grav/Mag/MT	Grav/Mag/MT	Grav/Mag/MT	Grav/Mag/MT			Microseis
	Fault delineation	FTG/MT	FTG/MT	FTG	FTG/Grav/Mag	FTG		
Play Fairway Analysis	Cross section modelling	Gravity/MT	Gravity/MT	FTG/MT	FTG/MT	FTG/CSEM	FTG/CSEM	

	Magnetics/MT	Magnetics/MT	Mag/CSEM/MT	Mag/MT/CSEM	Mag/MT/CSEM	Mag/MT/CSEM
Detection/delineation of volcanics						
Velocity constraints	Grav/Mag/FTG/MT	Grav/Mag/FTG/MT	FTG/MT/CSEM	FTG/MT/CSEM	FTG/OBN/CSEM FTG/OBN	FTG/OBN/CSEM FTG/OBN
PSDM modelling	Grav/FTG	Grav/FTG	Grav/FTG	FTG		
Seismic migration improvement	Grav/Mag. MT	FTG/MT	FTG/MT/CSEM	FTG/MT/CSEM		
Imaging prospect scale features	Grav/Mag/FTG/MT	FTG/MT	FTG/MT	CSEM	CSEM	OBN
Clastic/salt interface	Grav/Mag/MT	FTG/MT	FTG/MT	FTG/MT/CSEM	CSEM	CSEM
Clastic carbonate interface	Grav/Mag/MT	FTG/MT	FTG/MT	FTG/MT/CSEM	CSEM	CSEM
Prospect Analysis						
Prospect definition		FTG/CSEM	FTG/CSEM	FTG/CSEM	FTG/CSEM	OBN
PSDM modelling		FTG/CSEM	FTG/CSEM	FTG/CSEM	FTG/CSEM	FTG/CSEM
Prospect definition at relatively shallow depth (~2 km)		FTG/CSEM	FTG/CSEM	FTG/CSEM	FTG/CSEM	FTG/CSEM
Sub basalt imaging	FTG/MT/CSEM	FTG/MT/CSEM	FTG/MT/CSEM	FTG/CSEM	FTG/CSEM	FTG/CSEM
Sub Salt imaging	FTG/MT	FTG/MT	MT/CSEM	CSEM	OBN	OBN
Lithology prediction			FTG	FTG		OBN

Table 9.3 (*cont.*)

	Seismic coverage assumption	No 2D	2D	2D	2D/3D	3D	3D	3D
Production and Reservoir Management	Fluid prediction					CSEM	CSEM/OBN	CSEM/OBN
	Production monitoring							OBN/microseismic/CSEM, 4D and borehole gravity
	PSDM modelling						OBN	OBN
	Location of old wells and pipelines					Magnetics	Magnetics	Magnetics
	Fault location			Mag/FTG/Microseis	Mag/FTG/Microseis	Mag/FTG/Microseis	Mag/FTG/Microseis	Mag/FTG/Microseis
	Fault mobility			Microseis	Microseis	Microseis	Microseis	Microseis
	Fracture orientation			Microseis	Microseis	Microseis	Microseis	Microseis
	Fracture location							Microseis

CSEM, controlled-source electromagnetic; FTG, Full Tensor Gradiometry; Grav, gravimetry; Mag, magnetics; Microseis, microseismic; MT, magnetotelluric; OBN, ocean bottom node; PSDM, Pre-Stack Depth Migration.

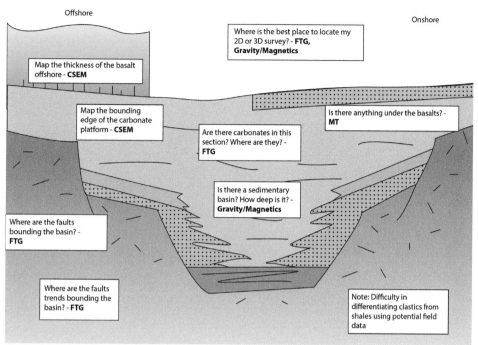

Figure 9.1 Geophysical technologies and the failed rift setting.

Figure 9.1 provides a graphic illustration of the use of special geophysical technologies in a failed rift setting. Note that potential fields technologies require density, resistivity or magnetic differences.

Figure 9.2 shows a similar analysis for a typical passive margin with the possible presence of carbonates, salt or volcanic.

Compressional terrains are normally onshore and associated with a difficult topography. Seismic is expensive. Airborne FTG, gravity and magnetic surveys, however, are fairly cheap and can add considerable value. EM techniques and in particular MT are particularly valuable in thrust settings and can be used to estimate the depth of individual thrust sheets, as illustrated in Figure 9.3

9.3 Integrated Interpretation

The theme running through this introduction is one of integration. In this section we have chosen three examples from the technology chapters to use as case studies to illustrate how these technologies can be combined with seismic to reduce subsurface risk.

OFFSHORE PASSIVE MARGIN SETTING

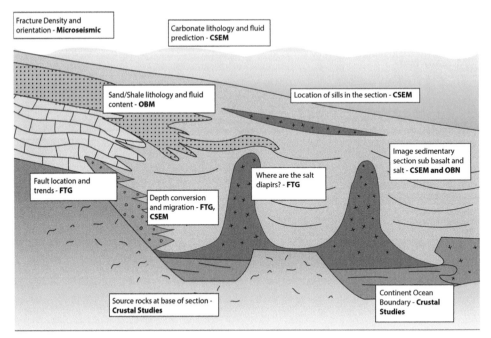

Figure 9.2 Geophysical technologies and the passive margin setting.

9.3.1 Nova Scotia: Magnetics, Crustal Seismic and Long-Offset 2D Seismic

Nova Scotia Offshore is an underexplored basin off eastern Canada. It is a classic passive margin whose basin history is dominated by two events:

1. Late Triassic continental rifting (leading to thick syn-rift salt deposits) followed by very early Jurassic breakup
2. Middle Jurassic to Early Cretaceous deltaics deposition into accommodation space created by basin subsidence, salt remobilisation and growth faulting

Historically hydrocarbon production has been from the Early Cretaceous deltaic deposits on the continental shelf. The continental slope was under explored. The Nova Scotia Department of Energy commissioned a basin scale Play Fairway Analysis to attract investment and stimulate exploration (Wilson et al., 2010).

The objective of the study was to investigate the possibility of a regional source rock that could open up the prospectivity of the deeper water. The syn-rift to early post rift section was the obvious starting point for the investigation. The classic hypothesis is that source rocks are likely to form in volcanic passive margin rift settings in which a restricted shallow marine setting is preserved for an extended

ONSHORE OVERTHRUST SETTING

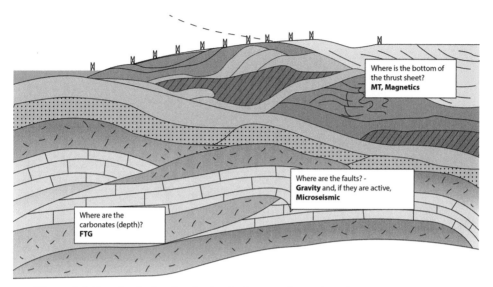

Figure 9.3 Geophysical technologies in the overthrust domain.

period of time immediately post rift. Therefore the challenge was to demonstrate that Nova Scotia/Morocco was a volcanic passive margin.

Magnetic data (Figure 9.4) show the presence of the East Coast Magnetic Anomaly (Wilson et al., 2010). Note the dark red (high magnetic anomaly) band running parallel to the Canadian coast. Note also that the band declines on moving North East along the margin. This was interpreted as implying that volcanics decline towards the east.

Figure 9.4 also shows the location of a long offset GXT 2D seismic grid. Seaward dipping reflectors, indicators of subaerial volcanism, are another line of evidence for a volcanic margin, were noted on GXT lines 1100 and 1400, the two western-most dip lines. These lines are through the heart of the magnetic anomaly. The long-offset 2D data was therefore integrated with the magnetic interpretation to search for evidence of volcanism.

Crustal structure is also an important indicator of the nature of rifting and can also provide heat flow calibration and the position of the Continent Ocean Boundary (COB).

A 400 km ocean bottom seismograph (OBS) line was acquired along the furthest North East GXT line 2000 on Figure 9.4. Figure 9.5 shows the results of the OBS line and illustrates the crustal structure that supported the concept of a volcanic margin.

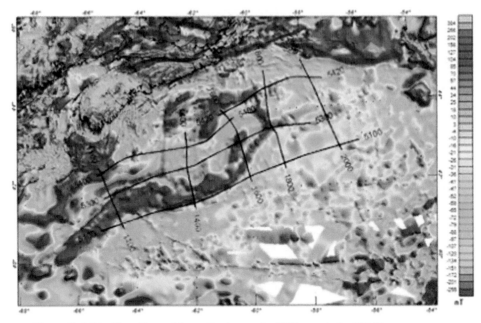

Figure 9.4 The East Coast Magnetic Anomaly (Wilson et al., 2010).

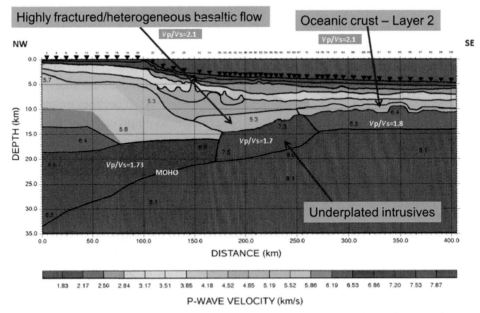

Figure 9.5 V_p velocity model of the Nova Scotia – North Atlantic passive margin. V_p/V_s values are shown in the model (Luheshi et al., 2012).

The high velocity layer marked in purple is interpreted as an underplated body thus indicating that the rift is volcanic in nature at this location.

The location of the OBS survey along the GXT line enabled a more detailed delineation of the COB and the definition of its seismic character. In addition, the OBS velocities were used to provide a more accurate depth conversion of the conventional seismic reflection data.

In parallel with the crustal studies work, a forensic geochemistry study was undertaken to understand the nature of the liquids found in the fields of Nova Scotia and their conjugates in Morocco. Further work considered the characteristics of source rocks found in well penetrations on both sides of the margin. For some of the oils we could demonstrate an explicit link with the penetrated source rocks, that is, type III terrigenous source rocks found in a deltaic setting. However, the biomarker analysis of the liquids found evidence of oils sourced from a marine hypersaline, stratified water column environment, inconsistent with a deltaic environment.

A likely setting in which such a depositional environment could exist is syn- or immediately post-rift in the transition from a constricted marine setting leading to halite deposition, to a more open marine setting as the continents drifted apart.

The geochemical analysis, combined with the crustal studies work on rifting mechanism provided powerful evidence for a new, as yet untested, lower Jurassic regional source rock offshore Nova Scotia and in its conjugate in Morocco. This project illustrates the power of integrating magnetic studies and conventional seismic and crustal seismic alongside geochemistry data to bring about transformation in perceptions on exploration potential in an unexplored frontier basin.

Exploration was re-invigorated in Nova Scotia with BP and Shell undertaking large exploration programmes. BP & Shell invested approximately $2 billion exploring dep water Nova Scotia. This included drilling three deep water exploration wells between 2014 and 2018, all of which failed to find commercial hydrocarbons. This work also contributed to a resurgence in exploration drilling offshore Morocco with a number of companies (including BP, Kosmos, Genel Energy, Cairn Energy and Chariot Oil & Gas) drilling deep water exploration wells between 2010 and 2020. This exploration campaign also failed to discover commercial hydrocarbons reserves.

9.3.2 Subsalt Imaging: Resolving the K-2 Salt Structure in the Gulf of Mexico

A classic integrated application of potential field and seismic data is provided by the example of the appraisal of the K-2 field, which lies some 175 miles SW of New Orleans in the Gulf of Mexico (Figure 9.6). The reservoir is of Miocene age

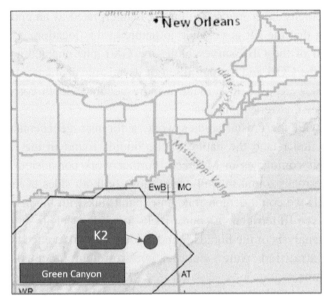

Figure 9.6 Location of K-2 field in the Gulf of Mexico.

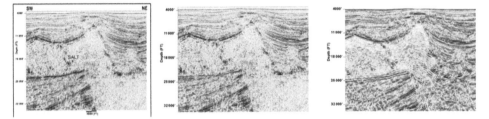

Figure 9.7 Pre-stack depth migration profile along line a through the K-2 field. (Left) Kirchhoff migration, showing poorly imaged area updip of the field. Centre: Kirchhoff migration with base of salt horizon in yellow, as determined by FTG inversion. This shows a symmetric salt keel. (Right) Wave-equation pre-stack depth migration, which also shows the presence of a salt keel; yellow horizon shows the FTG inversion result. (After O'Brien et al., 2005, figure 8)

and the trap is a three-way dip structure with a salt top seal. The field is estimated to hold some 100 mmbbls of oil and is at an average depth of some 25,000 feet subsea (water depth here is 4,326 feet). K-2 is operated by Anadarko Petroleum (now Occidental Oil Company), which holds 41.8% working interest in the field.

Appraisal of the reservoir was faced with a very significant uncertainty in the extent of the sands up dip beneath a thick salt structure. The image in Figure 9.7 shows a line from a pre-stack depth migrated 3D seismic survey. The left-hand panel shows the results of using Kirchhoff migration to process the data without

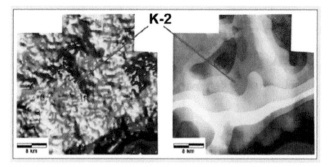

Figure 9.8 (Left) G_{zz} component of the gravity gradient field over K-2. (Right) Conventional 3D free air gravity over the same area. Note the superior definition of the gravity gradient data. (After O'Brien et al., 2005)

the FTG information: The poorly imaged area up dip of the field and the uncertainty in the termination of the reservoir against the overlying salt are clear.

The operator partnership was faced with two stark choices in managing this problem: either (1) drill an up dip well to test reservoir extent or (2) potentially leave a substantial amount of reserves undeveloped. The geophysical challenge was clear and conventionally the 'standard' approach would be to reprocess the 3D with newer/better imaging technology. As well as this, the operator decided to acquire a gravity gradiometry survey for two reasons: (1) to help with defining the velocity structure for a new depth migration and (2) to get an independent estimate of the base of salt morphology.

Imaging a salt/soft sediment interface using FTG is optimal from a density contrast perspective; however, the 25,000 feet depth of the base of salt was a significant challenge in terms of resolution.

An image of the G_{zz} data from the FTG survey is shown in Figure 9.8, which highlights the superior resolution of the FTG data.

Prior to the reprocessing and joint FTG inversion, there were three possible interpretations of the base of salt, namely (1) flat across the poor imaging area, (2) broad salt 'keel' and (3) a shallower base salt. The approach taken to address the problem was to acquire a marine FTG survey and to use this in a joint reprocessing project applying wave equation depth migration (the most advanced method routinely available in the early 2000s).

Figure 9.7 shows examples of an inline through the K-2 field. The first image is for the original depth migrated image showing the uncertainty in the morphology of the base of salt. The second image adds the depth of the base of salt as estimated from inversion of the FTG data and the final image shows the updated depth migration. The last version shows that the revised migration is in excellent agreement with the FTG result and hence gives confidence in the 'keel' interpretation of base of salt.

This result had a profound implication on the field development plan. Without this the operators would have had to chose between drilling an additional updip well or leaving some pay undeveloped.

9.3.3 Identifying Volcanics: CSEM and MT West of Shetlands

The application of marine EM west of Shetland to study basalt and sub-basalt structure has been a focus of attention for many years. In this environment understanding the thickness of any sedimentary sequences beneath the basalt is important in assessing prospectivity. Assuming a sedimentary sequence is present, understanding the structure of the overlying basalt is critical for the construction of accurate velocity models for seismic migration. There is often a large uncertainty on basalt thickness based solely on seismic data. Misinterpretations in excess of 1 km have been reported on the North Atlantic Margin (Hoversten et al., 2013). Such errors can have a significant impact on drilling plans and ultimate cost. Combining CSEM and MT data in this environment can allow basalt thickness, sub-basalt sediment thickness and basement structure to be estimated. Hoversten et al. (2013) present a calibration survey between two wells on the North Atlantic Margin that have penetrated beneath the basalt. By jointly inverting CSEM and MT data, both the thin basalt, underlying sedimentary sequences and resistive electrical basement are well resolved (Figure 9.9).

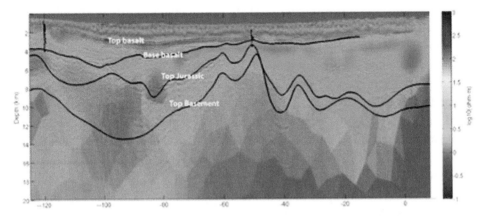

Figure 9.9 An example of sub-basalt structure constraint West of Shetland using the joint inversion of CSEM and MT data. In this case the top basalt boundary, which can usually be readily imaged with seismic, has been constrained by allowing a break in smoothness at this horizon. (From Hoversten et al., 2013)

This illustrates how using CSEM and MT in combination with conventional seismic can be used to provide a much more constrained solution.

References

Fraser, A. J. 2010. A regional overview of the exploration potential of the Middle East: A case study in the application of play fairway risk mapping techniques. In *Petroleum Geology Conference Proceedings*.

Grant, S., Milton, N. and Thompson, M. 1996. Play fairway analysis and risk mapping: An example using the Middle Jurassic Brent Group in the northern North Sea. *Norwegian Petroleum Society Special Publications*.

Hoversten, G. M., et al. 2013. CSEM & MMT base basalt imaging. In *75th European Association of Geoscientists and Engineers Conference and Exhibition 2013 Incorporating SPE EUROPEC 2013: Changing Frontiers*.

IEA (International Energy Agency). 2019. World Energy Outlook 2019: Analysis. Paris: International Energy Agency.

Karman, G., Ramirez, D., Voon, J. and Rosenquist, M., 2011. A decade of controlled-source electromagnetic, CSEM, in Shell: Lessons from a global look back study. Presented at the 4th NPF Biennial Petroleum Geology Conference, Bergen.

Levell, B., Argent, J., Doré, A. G. and Fraser S. 2010. Passive margins: Overview. In *Petroleum Geology Conference Proceedings*.

Nemcok, M. 2016. Models of source rock distribution, maturation, and expulsion in rift and passive margin settings. In *Rifts and Passive Margins*, 347–75. Cambridge: Cambridge University Press.

Roberts, D. and Bally, A. 2012. Regional geology and tectonics: Phanerozoic passive margins, cratonic basins and global tectonic maps. In Roberts, D. G. and Bally, A. W. (eds.), *Regional Geology and Tectonics: Phanerozoic Passive Margins, Cratonic Basins and Global Tectonic Maps*. Amsterdam: Elsevier.

United Nations. 2015. About the Sustainable Development Goals – United Nations Sustainable Development. *Sustainable Development Goals*.

White, R S. et al. 1987. Magmatism at rifted continental margins. *Nature*, **330**, 439–44.

Wilson, H. A.M., Luheshi, M. N., Roberts, D. G. and MacMullin, R. A. 2010. Play fairway analysis, offshore Nova Scotia. In *Proceedings of the Annual Offshore Technology Conference*.

Glossary

Term	Definition
1-C	single-component
3-C	three-component
AGG	Airborne Gravity Gradiometry
Basement	deepest geological layer, usually Precambrian igneous or metamorphic rock, usually of no direct interest in petroleum exploration
Bboe	billions of barrels oil equivalent
BGI	International Gravity Bureau
BGS	British Geological Survey
BIF	Banded Iron Formation (is V strongly magnetic)
BMR	Bureau of Mineral Resources – precursor to Geoscience Australia
Bopd	Barrels of oil per day
BRGM	Geological Survey of France
BSR	Bottom Simulating Reflector
cgs-emu	electromagnetic units in the centimetre-gram-second system of units
COB	Continent/Ocean Boundary
CSEM	Controlled Source EM – a survey method using an artificial EM source
Curie Point Isotherm	depth below which rocks are not ferromagnetic, by virtue of temperature
DAS	Distributed Acoustic Sensing
DC	direct current
Deconvolution	computer process that attempts to recover geological structure from geophysical observations
DFN	Discrete Fracture Network

DFIT	Diagnostic Fracture Injection Test
DLR	German Aerospace Centre
DTS	Distributed Temperature Sensing
EM	electromagnetic
ESA	European Space Agency
EUR	Estimated Ultimate Recovery
FDI	Fracture-Driven Interaction
FFT	Fast Fourier Transform
FoM	figure of merit – specification (from GSC) of Aircraft Manoeuvre Magnetic Noise
FTG	Full Tensor Gradiometry
Gal	unit of gravity anomaly (acceleration). $1 \text{ cm} \cdot \text{s}^{-2}$
GEBCO	General Bathymetric Chart of the Oceans
Geoid	Gravity equipotential surface coinciding with sea level
GLONASS	Russian satellite navigation system
GOCE	Gravity Field and Steady-State Ocean Circulation Explorer – an ESA satellite programme to measure gravity in orbit
GPS	Global Positioning System - US satellite navigation system
GRACE	Gravity Recovery and Climate Experiment a NASA/DLR satellite system to detect gravity changes in orbit
GSC	Geological Survey of Canada
HED	Horizontal Electric Dipole – a CSEM source type
HRAM	High-Resolution Aeromagnetic
IGRF	International Geomagnetic Reference Field
IGSN	International Gravity Standard Network
IPHT	Leibniz Institute for Photonic Technology, Jena, Germany
IPR	highest Initial Production Rate
LF	low frequency
LIDAR	Light Detection And Ranging
LOFS	Life of Field Seismic
Maxwell's equations	system of equations describing all electromagnetic phenomena
MEMS	microelectromechanical systems
Moho	Mohorovičić Discontinuity – seismic velocity boundary taken to be the base of the earth's crust
MT	magnetotelluric – a survey method using natural EM fields as a source
MTI	Moment Tensor Inversion

NASA	National Aeronautics and Space Administration – an agency of the US Government
NOAA	HYPERLINK "https://www.noaa.gov/" National Oceanic and Atmospheric Administration – an agency of the US Government
PRM	Permanent Reservoir Monitoring
PSDM	Pre-Stack Depth Migration – an intensive seismic data processing procedure
ROV	remotely operated vehicle
SDM	Source Dipole Moment – a measure of the strength of a CSEM source
SI	International System of Units, the internationally agreed implementation of the metre–kilogramme–second system of units
SQUID	Superconducting Quantum Interference Device – a very sensitive magnetometer
SRV	Stimulated Reservoir Volume
TTI	Tilted Transverse Isotropy
UCSD	University of California at San Diego
USGS	United States Geological Survey
VED	Vertical Electric Dipole – a CSEM source type
VTI	Vertical Transverse Isotropy
WARRS	Wide Aperture Reflection and Refraction Seismics
WWII	World War Two

Index

4D, 226
    active seismic imaging, 283
    surveys, 250
accelerations due to the moving platform., 128
acoustic impedance, 43
acquisition, 223, 229–30, 251, 257
acquisition rates, 226
additional information provided by the tensor
    gradients, 137
airborne gravity acquisition, 125
airgun, 220, 227
    sources, 267
aliasing, 141
ambient tow noise, 240
amplitude variation with offset (AVO), 240
Angola, 11
anisotropy, 286
    electrical, 172–4, 185
apparent velocity, 42
appraisal, marine EM, 207–8
ARKeX, 128
arrival time picking, 285
Atlantis Field, 261
Austin Bridgeporth, 130
autonomous nodes, 263
autonomous underwater vehicle (AUV), 224
    nodes, 251
azimuth, 223, 227, 230, 232
    'FAZ', 236
azimuthal coverage, 233

bandwidth, 222, 227–8, 232, 239
Barents Sea, marine EM, 199–205
Barnett Formation, 273
basin analysis, 8
basin floor turbidites, 18
basin morphology, 16
basin screening, 8
battery life, 266
beach ball, 291

beam steering, 285
Bell Aerospace, 128
Bell Geospace, 128
BHP-Billiton, 128
blended, 266
blended source, 257
blending, 257
broadband, 231
    seismic, 231
Broadseis^{TM}, 231
Brune model, 290
Burgan, 73
buried array, 279

carbon capture and storage, 159
    monitoring, marine EM, 210
carbon capture, usage and sequestration (CCUS),
    210
case studies, 141
    gravity inversion, 110
    Guinea offshore, 101
    integrated modelling, 111
    Lake Edward, 103
    sedimentary structure, 106
CGG Airborne, 128
Clock drift, 269
CO_2-injection, 269
CO_2 sequestration, 306
coil, 236, 306
    shooting, 236
Coil^{TM}, 230, 236
collapse structure, 230
commercial application, 127
common receiver gathers, 40
common risk segment, 15
common shot gathers, 40
compressional wave, 39
compressive stress, 293
Conrad, 39
continental crust, 55

continent–ocean boundary, 49
continuous line acquisition, 49
controlled source EM method, 176–7
converted shear waves, 258
corner frequency, 290
cost, 66
    marine EM, 66
creaming curves, 10
critical angle, 41
CSEM. *See* Controlled Source EM method
Curie point isotherm, 86

data processing, 134, 226, 229, 231, 257
de-blending, 257
deconvolutions and depth estimators, 98
deep seismic sounding, 60
density, 39
deployment, 226, 243–4, 249
    rates, 245–6
depth estimation
    Euler, 98
    GridSLUTH, 98
    Naudy, 98
    spectral slope, 98
    SPI, 98
    Werner, 98
discrete fracture network, 299
displacement, 290
distributed acoustic sensing, 284
distributed temperature sensing, 305
diving wave, 45, 305
dots-in-the-box, 289
downhole arrays, 273
drainage network, 298
drainage volume, 290
dual sensor, 232
dual-sensor recording, 240
dynamic model, 21

East African Rift, 147
East Coast Magnetic Anomaly, 57
East European Craton, 60
Encana Blacksmith Buried Array, 281
enhanced Full Tensor Gradiometry (eFTG), 281
enhanced gravity system, 130
enhanced oil recovery', 21
eötvös, 126, 131
equipment, 49
event magnitude, 273
expected ultimate recovery, 289
exploration plays, 10
Exploration Triangle, 15
exploration, marine EM, 207
extensional tectonic environment, 17

Falcon™, 128
Faroe Shetland trough, marine EM, 194–5

Faroe-Shetland Basin, 152
Faust, 43
feasibility models, 136
fibre optic, 268
    cables, 304
finding cost per barrel, 9
first arrivals, 45
first breaks, 45
'fizz gas' problem, 170, 200
flexible azimuth sampling, 232
flying nodes, 250
focal mechanism, 276, 290–1
focal sphere, 278
foreland basins, 17
frac height, 288
frac hits, 305
frac length, 288
frac'ing', 272
fracture area, 290
fracture-driven interaction, 295
fracture plane, 294, 304
free falling OBS, 52
FreeCable™, 238
freshwater aquifer mapping, marine EM, 211
frontier exploration, marine EM, 206–7
FTG using moving platforms, 126
full azimuth (FAZ), 126
Full Tensor Gravity (FTG), 125
full waveform inversion (FWI), 45, 220
full waveform migration, 285, 287

Gabon salt and carbonate imaging, 143
Gardner, 44
gas hydrate mapping, marine EM, 211
geological model, 43, 49, 242, 250
geophones, 263
Ghana, 11
Ghawar, 73
global gravity models, 77
    EGM2008, 78
    GOCE, 78
global magnetic model, EMAG2, 78
GMA systems, 134
GRACE global gravity models, 77
gravity
    Bouguer and Terrain correction, 89
    Eötvös correction, 89
    free air correction, 89
    gravity field units, 88
    isostatic correction, 90
    isostatic residual, 102
    standard corrections, 89
    survey design, 92
gravity gradient imaging (GGI), 126
gravity gradiometry, 125
gravity instruments, 74
Greenland Passive Margin, 153

gross depositional environment, 15
Gulf of Mexico, marine EM, 197–8
gyroscopically stabilised platforms, 125

heat flow, 17
HED. *See* horizontal electric dipole
Hellenic arc, 62
historical production and costs, 132
horizontal electric dipole CSEM method, 177–9
horizontal well, 277
hot margins, 18
hydraulic fracturing, 272, 305
hydrophone, 49, 227, 230–1, 241, 244, 263
hypocentre, 272, 285
    estimates, 285

imaging, 228
Initial Production rate, 299
instruments, 49
integration, 65
interface, 41
interpolating faults seen on regional 2D seismic, 143
interpretation, 137
    marine EM, 187–90
interpretive enhancements, 17
intra-cratonic basin, 17
inversion, marine EM, 187–9

joint inversion, 65, 191, 193, 196–200, 217

K-2 salt structure in the Gulf of Mexico, 144
Kenya, 54

LIDAR, 134
life-of-field, 224
    seismic, 313
lithology, 43
Lockheed Martin, 131
    FTG, 239
low cable noise, 131

magnetic field
    International Geomagnetic Reference Field (IGRF), 94
        standard corrections, 95
        survey design, 96
        units, 93
magnetometers, 75
magnetotelluric method, 175–6
massive sulphide deposit mapping, marine EM, 211
material balance approach, 299
maximum horizontal stress, 288
megasequence, 17
Microelectromechanical System MEMS, 242, 244, 250
microseismic technology, 272
migration, 50
military application, 128

minimum horizontal stresses, 303
modelling, marine EM, 187–8
Mohorovicic (Moho) Discontinuity, 39, 187–8
moment magnitude, 290
moment tensor, 276, 293
    inversion, 276, 293
monitor well, 275
monitoring, marine EM, 208
MT. *See* magnetotelluric method
multi-azimuth (MAZ), 224, 234
    techniques, 224, 234
multichannel seismic (MCS), 58
multiphysics analysis
    reservoir characterisation, 190, 199–205
    structural, 190–9
multiple, 230
    suppression, 239
multiples, 240
multi-streamer, 234, 244

narrow azimuth ("NAZ"), 230
Nash Salt Dome, 71
natural source EM method. *See* magnetotelluric
    method
Net Present Value, 301
Neuquén Basin, 294, 306
nodal CSEM, 177–8
node, 245
    deployment, 245
    design and deployment, 245
    positioning, 269
node-on-node 4D, 265
nodes-on-ropes, 246
noise, 134
    spectra, 241
normal incidence, 40
Nova Scotia, 48

ocean bottom, 220
    seismic, 224, 242, 254
    seismograph, 243
    seismometer, 243
ocean bottom cable (OBC), 220, 224, 239, 258
    geophone, 243
ocean botom node (OBN), 51
    data, 51
ocean bottom seismometer (OBS), 51
oceanic crust, 51
opportunity funnel, 12
optical, 254
optical distribution unit, 268
optical fibres, 224, 268
OptoSeis™, 254
Orion™, 230

passive margin, 11
    basins, 17

Pau reservoir monitoring, 160
permanent reservoir monitoring (PRM), 224
    optical fibre, 313
    systems, 254
Permian Basin, 300
permitting, 274
petroleum system, 8
petrophysical joint inversion, 191
Pg, 42
physical properties
    density, *82*
    Gardner's relation, 83
    magnetic susceptibility and remanence, 84
plate tectonics, 16, 67
platform accelerations, 125
platforms, 220, 228, 241
play fairway analysis, 8
PmP, 42
Pn, 42
Poisson's ratio, 44, 170, 200
potential field methods, 124
pre-stack depth migration, 145
pre-survey modelling, marine EM, 185–6
principal stresses, 304
    components, 302
processing of gravity gradiometry, 136
productive SRV, 299
proppant, 289, 299
prospect definition, 8
prospect ranking, marine EM, 207
P-wave, 39

radiation pattern, 290
rake, 302
rate of subsidence, 18
ray-paths, 223
ray-tracing, 46
recoverability, marine EM, 185
Red Sea, 147
    marine EM, 198
'reduced' travel time–distance plots, 42
reduction velocity, 42
reflection coefficient, 43
regional compilations, *75*
regional play analysis, marine EM, 206–7
regional screening, 7
relative energy, 43
remotely operated vehicles (ROV), 51
    deployment, 51
repeatability for 4D surveys, 250
reservoir, 21
    compaction, 310
    monitoring, 310
    monitoring and CCS, 310
    simulation, 310
resistivity, 172–3

resolution, 221, 229, 231, 233
    of airborne gravity surveys, 126
    marine EM, 188–90
return on investment, 299
rich azimuth (RAZ), 236
    shooting, 236
rich, full or all azimuth, 224
robotics, 250
rock bursts, 306
rock physics, 306
rock rigidity, 290
Rocky Mountain Arsenal, 272
rotating wheel type, 131

San Andres formation, 300
satellite, 76
    geodesy, 76
    navigation, 76
scalar potential, 129
sea floor subsidence, 310
seaward dipping reflectors, 18
seismic potency, 290
seismic quality, 227
sensitivity, 126
    marine EM, 181–4
sequence structure, 19
shale gale, 273
shear, 304
    failure, 304
    stress, 304
    wave, 304
    wave aeismology, 304
shear-wave, 43
Sigsbee Escarpment, 263
simultaneous source, 220
single azimuth, 234
Sirte Basin, 73
Sleipner field, 161
source rock, 17
spatial resolution, 126
spatial sampling, 230
stage spacing, 288
staged approach, 7
static model, 21
stimulated reservoir volume, 288
stimulation, 291
streamer, 230, 232
    construction, 227
    steering, 234
stress and geomechanics, 302
stress tensor, 303
strike-slip, 291
structural aliasing, 126
structural correlation, 141
sub-basalt imaging, 152
    marine EM, 152

subsalt, 222, 261
  imaging, 157
  imaging, marine EM, 167, 170
surface array, 278
surface ghost, 228
surface-related multiple contamination, 261
survey design, 135
S-wave, 41
synthetic gather, 47

target suitability, marine EM, 185–7
temporal bandwidth, 229
tensile stress, 293
tensors, 130
terrain and bathymetry
  GEBCO, 86
  SRTM, 86
three components, 128
thrusted compressional basins, 17
time-lapse ('4D'), 250
tomography, 41
towed cable, 223
towed streamer, 224, 239
  2D, 224
  3D, 226
  CSEM, 178–9, 200
towed-cable, 220

UK/Norway, 11
units, 44
US Gulf of Mexico, 11

Vaca Muerta, 306
  Formation, 294

value, 140
  of finer sampling, 140
  of information, 260
  of seismic data, 260
V_C, 42
VED. *See* vertical electric dipole CSEM method
velocity depth model, 45
velocity gradient, 44
velocity model, 285
  for seismic imaging, 126
vertical electric dipole CSEM method, 179, 174, 180
via full waveform inversion, 179, 220
Vp, 39, 332
Vp/Vs, 44, 332
Vs, 41
VSPs, 268, 269

Water injection, 266
wavefield, 227, 229, 231, 241
well productivity, 20
well spacing, 288
wellbore damage, 293
West Africa, 11
wide-angle towed streamer (WATS), 224, 234
wide aperture, 234
Wide Aperture Reflection and Refraction Seismics (WARRS), 130
wide azimuth (WAZ), 40
wide-angle, 224
  towed streamer, 224
wide band gravity, 224

Zoeppritz, 41